Dependable Computing and Fault-Tolerant Systems

Edited by
A. Avižienis, H. Kopetz, J. C. Laprie

Advisory Board
J. A. Abraham, V. K. Agarwal, T. Anderson, W. C. Carter,
A. Costes, F. Cristian, M. Dal Cin, K. E. Forward, G. C. Gilley,
J. Goldberg, A. Goyal, H. Ihara, R. K. Iyer, J. P. Kelly,
G. Le Lann, B. Littlewood, J. F. Meyer, B. Randell,
A. S. Robinson, R. D. Schlichting, L. Simoncini, B. Smith,
L. Strigini, Y. Tohma, U. Voges, Y. W. Yang

Volume 7

Springer-Verlag Wien New York

H. Kopetz, Y. Kakuda (eds.)

Responsive Computer Systems

Springer-Verlag Wien New York

Prof. Hermann Kopetz, Technische Universität Wien, Austria

Dr. Yoshiaki Kakuda, Osaka University, Japan

With 96 Figures

ISSN 0932-5581
ISBN-13: 978-3-7091-9290-0   e-ISBN-13: 978-3-7091-9288-7
DOI: 10.1007/978-3-7091-9288-7

FOREWORD

For the second time the International Workshop on Responsive Computer Systems has brought together a group of international experts from the fields of real-time computing, distributed computing, and fault-tolerant systems. The two day workshop met at the splendid facilities at the KDD Research and Development Laboratories at Kamifukuoka, Saitama, in Japan on October 1 and 2, 1992. The program included a keynote address, a panel discussion and, in addition to the opening and closing session, six sessions of submitted presentations.

The keynote address "The Concepts and Technologies of Dependable and Real-time Computer Systems for Shinkansen Train Control" covered the architecture of the computer control system behind a very responsive, i.e., timely and reliable, transport system–the Shinkansen Train. It has been fascinating to listen to the operational experience with a large fault-tolerant computer application. "What are the Key Paradigms in the Integration of Timeliness and Reliability?" was the topic of the lively panel discussion. Once again the pro's and con's of the time-triggered versus the event-triggered paradigm in the design of a real-time systems were discussed. The eighteen submitted presentations covered diverse topics about important issues in the design of responsive systems and a session on progress reports about leading edge research projects. Lively discussions characterized both days of the meeting.

This volume contains the revised presentations that incorporate some of the discussions that occurred during the meeting. Since it is hardly possible to record the engaged discussions without interrupting the flow of the arguments during the event, the interested reader is invited to attend the Third Workshop on Responsive Systems, to be held in New Hampshire from September 28 to October 1, 1993, if he/she wants to get the full spirit of the meeting.

At this point we would like to express our gratitude to the members of the program committee who have carefully reviewed the submitted papers. We would also like to thank the General Chairman, Tohru Kikuno and Mirsolaw Malek for their expert guidance and dedicated support of the Program Committee. If the work can be measured by the number of e-mail interactions during the preparation of this meeting, here is a short statistics. There exchanged 161 messages between University of Texas at Austin and Osaka University for organization of the workshop and 75 messages between Technical University of Vienna and Osaka University for determination of the technical program.

Last but not least, special thanks are due to staff members of *KDD R&D* Laboratories, especially Hironori Saito, for their endless efforts and dedication to making the workshop successful, Yukio Mohri, *NihonUnisysLtd.* for his outstanding job in the production of the workshop proceedings, and Hideki Yukitomo, a graduate student of Osaka University, for his excellent work in reformatting the papers for this volume.

Hermann Kopetz   Yoshiaki Kakuda
Program Co-Chair   Program Co-Chair
Technical University of Vienna Osaka University
Austria   Japan

Sponsors

IEEE Computer Society Technical Committee on *Fault-Tolerant Computing*
IEEE Computer Society Technical Committee on *Real-Time Systems*
IEICE Technical Group on *Fault-Torerant Systems*
International Communications Foundation
International Information Science Foundation
Kokusai Denshin Denwa Co., Ltd. (KDD)
U. S. Office of Naval Research

Conference Organization

General Co-Chairmen

Miroslaw Malek
University of Texas at Austin
USA

Tohru Kikuno
Osaka University
Japan

Program Co-Chairmen

Hermann Kopetz
Technical University of Vienna
Austria

Yoshiaki Kakuda
Osaka University
Japan

Session Chairs

Donald Fussell
University of Texas at Austin
Austin, Texas, USA

Haruhisa Ichikawa
NTT
Japan

Yoshiaki Kakuda
Osaka University
Toyonaka, Osaka, Japan

Gary Koob
ONR
USA

Hermann Kopetz
Technical University of Vienna
Vienna, Austria

Insup Lee
University of Pennsylvania
USA

Miroslaw Malek
University of Texas at Austin
Austin, Texas, USA

Kinji Mori
Hitachi Co.
Japan

Program Committee

Tom Anderson
University of Newcastle
UK

Flaviu Cristian
University of California at San Diego
USA

Haruhisa Ichikawa
NTT
Japan

Gerard Le Lann
INRIA
France

Gary Koob
ONR
USA

Al Mok
University of Texas at Austin
USA

Sachio Naito
Tokyo Metropolitan University
Japan

Fabio Panzieri
University of Bologna
Italy

David Powell
LAAS-CNRS
France

Krithi Ramamritham
University of Massachussetts
USA

Lui Sha
Carnegie Mellon University
USA

Yoshihiro Tohma
Tokyo Institute of Technology
Japan

Mario Tokoro
Keio University
Japan

Yoshiyori Urano
KDD
Japan

Paulo Verissimo
INESC
Portugal

Contents

Opening Session

Six Difficult Problems in the Design of Responsive Systems

H. KOPETZ

Institut für Technische Informatik

Technische Universität Wien

Treitlstr. 3/182

A-1040 Wien, Austria

email:

hk@vmars.tuwien.ac.at

Abstract

Any responsive system has to provide solutions to the following six difficult problems: flow control, scheduling, testing for timeliness, timely error detection, replica determinism, and redundancy management. The first three problems have to be addressed in any real-time system that guarantees deadlines. The second set of problems must be solved if fault-tolerance is to be implemented by active redundancy. This paper examines the characteristics of these problems and evaluates solutions in event-triggered and time-triggered architectures.

Key Words: Responsive Systems, Real-time Systems, Fault Tolerance, Distributed Systems

1 Introduction

A responsive system is a distributed fault-tolerant real-time system [1]. It has to meet the deadlines dictated by its environment, otherwise it has failed. In this paper we focus on those applications, where such a failure can have catastrophic consequences, e.g. a flight control system, a drive by wire systems, or a nuclear power control system. In these applications a high level of assurance concerning the proper operation of the system–both in the domain of values and in the domain of time–is demanded. Designing responsive systems is challenging: the functional specifications must be met within the specified deadlines. The strict observance of these deadlines complicates the solutions of many problems that are well understood if no time constraints have to be watched. In this paper we analyze six fundamental problems of this kind: flow control, scheduling, testing for timeliness, timely error detection, replica determinism, and redundancy management. We further investigate how different architectures cope with these problems. For our purpose it is useful to classify an architecture as either even-triggered (ET) or time-triggered (TT). In an ET architecture a system activity is initiated as an immediate consequence of the occurrence of a significant event in the environment or the computer system. A TT architectures observes the state of the environment periodically and initiates system activities at recurring predetermined points in the globally synchronized time [2]. Contrary to the prevailing practice in real-time system design, the listed problems are more difficult to solve in ET architectures, such as DELTA4 [3], MAFT [4] than in TT architectures such as MARS [5]. This paper is organized as follows. In the next section the architectural assumptions are presented and a comprehensive description of the properties of ET and TT architectures is given. In the following section, the core section of this paper, the six problems are explored in detail and possible solutions in TT and ET architectures are examined. The results of this analysis are put into perspective in relation to the prevailing practice in the design of real-time systems.

2   Architectural Assumptions

A real-time application can be decomposed into a controlled object (e.g., the machine that is to be controlled) and the controlling computer system. The computer system has to react to stimuli from the controlled object within an interval of real-time that is dictated by the dynamics of the controlled object. We assume that the computer system is distributed. It consists of a set of self-contained computers, the nodes, which communicate via a real-time Local Area Network (LAN) by the exchange of messages only. Some nodes, the interface nodes, support a connection to the intelligent instrumentation, i.e., the sensors and actuators, which are at the interface to the controlled object. In a real-time

LAN the maximum delay of a message exchange, i.e., the execution of the communication protocol, has to be bounded by a time interval d_{max}. The difference between this maximum delay d_{max} and the minimum delay $d_{min}\epsilon = d_{max} - d_{min}$ is called the temporal uncertainty ϵ of the protocol. In responsive systems, both d_{max} and ϵ have to be small [6]. From the point of view of the control system designer, the behavior of a real-time application can be modelled by a set of computational processes operating on representations of RT-entities. A RT-entity is an element of interest for the given purpose, either in the controlled object or in the computer system. A RT-entity has a time-dependent internal state. Examples of RT-entities are the temperature of a particular vessel (external RT-entity), the speed of a vehicle (external RT-entity), or the intended position of a specific control valve (internal RT-entity). The observation of a RT-entity is stored in a RT-object in the computer system.

2.1 Event-Triggered Architectures

In a purely event triggered architecture all system activities are initiated by the occurrence of significant events in a RT-entity or a RT-object. A significant event is a change of state in a RT-entity or RT-object which has to be handled by the computer system. Other state changes of RT-entities are considered insignificant and are neglected. In many implementations of ET-systems the signalling of significant events is realized by the well known interrupt mechanism, which brings the occurrence of a significant event to the attention of the CPU. A significant event in the environment, noticed by an interface node, will most often cause a dynamic scheduling decision to activate the appropriate processing task and the transmission of a message to the other nodes in the system. To avoid a congestion of the communication system in case of coincident events, explicit congestion control mechanisms have to be implemented in the network layer of the communication system. Queues have to be provided in the operating system for the required intermediate storage of these events. To maintain a consistent order of the events at all receivers, atomic broadcast [7, 3] protocols have been proposed as basic communication mechanisms in distributed ET-systems. The temporal uncertainty of these asynchronous communication protocols can be significant and has an adverse effect on the temporal accuracy of the information [6].

2.2   Time-Triggered Architectures

Whereas in an ET-system the information about the occurrences of
events has to be disseminated promptly within the distributed computing system, the rapid diffusion of the observed states of the RT-entities
to all nodes of the distributed computing system is the fundamental
concern in a TT-system. This dissemination of the states is performed
periodically with periods that match the dynamics of the relevant RT-
entities. We call the periodic sequence of time points at which an RT-
entity is observed its observation grid [8]. Independently of the current
activity in a RT-entity, the state of a RT-entity is sampled at a constant rate. Although it is impossible to influence the time of occurrence
of a event in the controlled object, the time of recognition of such an
event, i.e. the polling of an event occurrence, is restricted to the grid
points of the observation grid in a TT-architecture. In a TT-system it
is the duty of the instrumentation to store an event occurrence until the
next polling cycle. Because of this enforced regularity, TT-architectures
are inherently more predictable than ET-architectures at the expense
of a more elaborate instrumentation and of a potential additional delay of one cycle for polled events (but not for sampled states). The
most common message handling primitives in distributed systems are
biased toward event information, i.e., a new message is queued at the
receiver and consumed on reading. In TT-systems a more appropriate
message model for handling state information is required. We call such
a message model a state message model and the corresponding message
a state message [9]. The semantics of a state message is related to the
semantics of a variable in a programming language. A new version of a
state message overwrites the previous version. A state message is not
consumed on reading. State messages are produced periodically at predetermined points in real-time, the grid points of the observation grid,
known a priori to all communicating partners. The information flow in
TT-protocols is unidirectional. It is the responsibility of the receiver to
verify that all required messages are available at the proper time.

3 The Six Problems

3.1 Flow Control

Flow control is concerned with the synchronization of the speed of the sender of information with the speed of the receiver, such that the receiver can follow the sender. Since the controlled object is usually not in the sphere of control of the computer system, there must be mechanisms that shield the computer system from overload scenarios. The definition of a proper flow control schema that will protect the computer system from overload, e.g., that caused by sensor failures, is one of the most difficult problems during the design of a responsive computer system.

Event-Triggered Architectures

Since an ET-system is demand driven, explicit bidirectional flow control mechanisms with buffering have to be implemented between a sending and a receiving node. Provisions must be made that event showers that originate in the controlled object can be buffered close to the interface of the computer system. Sometimes it may even be decided to discard events which occur too frequently. Several ad hoc engineering solutions are applied to restrict the flow of events at this interface. These include hardware implemented low pass filters, intermediate buffering of events in hardware and/or software, etc.. The time span which an event message has to wait in a buffer before it can be processed reduces the temporal accuracy of the observation and must thus be limited. The estimation of the proper buffer sizes is a delicate problem in the design of ET-systems.

Time-Triggered Architectures

During system design appropriate observation-, message-, and task-activation rates are fixed for the different RT-entities, based on their specified dynamics. The flow control in a TT-system is thus realized by an implicit mechanism. It has to be assured at design that all receiver processes can handle these rates. If the RT-entities change faster than specified, then some short lived intermediate states of the RT-entities will not be reflected in the observations and will be lost. Yet, even in a peak load situation, the number of messages per unit time, i.e., the message rate, remains constant. This implicit flow control will only function properly if the instrumentation of a TT-system supports the state view, i.e., the information produced by a sensor must express the current state of the RT-entity and not its change of state. If necessary, a local microcontroller has to transform the initial event information generated by an event sensor into its state equivalent. Consider the example of a push button, which is a typical event sensor. The local logic in this sensor must assure that the state "push button pressed" is true for an interval that is longer than the grid granularity of the observation grid of this RT-entity.

3.2   Scheduling

Scheduling is concerned with determining which task is to be executed next and what resources should be allocated to each task. We distinguish between two scheduling strategies, dynamic and static scheduling. In dynamic scheduling the scheduling decisions are made on-line on the basis of the current requests. In static scheduling, the scheduling decisions are made off-line and recorded in a dispatcher table. During run time a table lookup is performed by the dispatcher.

Event-Triggered Architectures

Operating systems for ET-systems are demand driven and require a dynamic scheduling strategy. In general, the problem of deciding whether a set of real-time tasks, the execution of which is constrained by dependency relations (e.g., mutual exclusion, precedence and pipelining), is schedulable is NP-hard [10]. Finding a feasible schedule, provided it exists, is another difficult problem. The known analytical solutions to the dynamic scheduling problem [11, 12] assume stringent constraints on the task set. At present it is not known how to derive realistic analytical guarantees for the deadlines of a dynamic distributed system. In practice most ET-systems resort to static priority scheduling. During the commissioning of the system the static priorities are tuned to handle the observed load patterns. No analytical guarantees about the peak load performance can be given.

Time-Triggered Architectures

Operating systems for TT-systems are based on a set of static predetermined schedules, one for each operational mode of the system. These schedules have to contain all necessary task dependencies and provide an implicit synchronization of the tasks at run time. They are constructed at compile time. At run time a simple table lookup is performed by the operating system to determine which task has to be executed at particular points in real time, the grid points of the action grid [8]. The difference between two adjacent grid points of the action grid determines the basic cycle time of the real-time executive. The basic cycle time is a lower bound for the responsiveness of a TT-system. In TT-systems all input/output activities are preplanned and realized by polling the appropriate sensors and actuators at the specified times. Access to the LAN is also predetermined, e.g., by a synchronous time division multiple access protocol. In the MARS system [5], which is a TT-architecture, the gridpoints of the observation grid, the action grid, and the access to the LAN are all synchronized with the global time. Since the temporal uncertainty of the communication protocols is smaller than the basic cycle time, the whole system can be viewed as a distributed state machine [13].

3.3   Testing for Timeliness

At present, 50% and more of the resources spent during the design and implementation of a major responsive system project are allocated to the testing phase. There is hardly any systematic methodology for testing the timeliness of real-time system.

Event-Triggered Architectures

In a system where all scheduling decisions concerning the task execution and the access to the communication system are dynamic, no temporal encapsulation exists, i.e., a variation in the timing of any task can have consequences on the timing of many other tasks in different nodes. Therefore it is not possible to perform systematic timing tests at the task or node level. Testing for timeliness has to be performed on simulated loads at the system level. Testing on real loads is not sufficient, because the rare events, which the system has to handle (e.g., the occurrence of a serious fault in the controlled object), will not occur frequently enough in an operational environment to gain confidence in the peak load performance of the system. The predictable behavior of the system in rare-event situations is of paramount utility in many real-time applications. The critical issue during the evaluation of an ET-system is thus reduced to the question, whether the simulated load patterns used in the system test are representative of the load patterns that will develop in the real application context. This question is very difficult to answer with confidence. Field maintenance data indicate that some application scenarios cannot be adequately reproduced in the simulated system test [14].

Time-Triggered Architectures

TT-systems require careful planning during the design phase. Because of this accurate planning effort, detailed plans for the temporal behavior of each task and access to the communication medium are available and the intended behavior of the system in the domain of time can be predicted precisely. Since the time-base is discrete and determined by the granularity of the action grid, every input case can be observed and reproduced in the domains of time and value. TT-systems enforce the temporal encapsulation of tasks and support a constructive test methodology. To achieve the same test coverage, the effort to test an ET-system is much greater than that required for the testing of the corresponding TT-system. The difference is caused by the smaller number of possible execution scenarios that have to be considered in a TT-system, since in such a system the order of state changes within a granule of the observation grid is not relevant [15].

3.4   Timely Error Detection

A responsive computer system must provide the specified service, even if faults covered by the fault-hypothesis, occur in some if its subsystems. The detection and passivation of errors must be performed before the consequences of the error impact the controlled object. If a node of the distributed computer system can exhibit an unconstrained behavior in case of an error, complex time-consuming agreement protocols are necessary at the architectural level to diagnose and mask node failures. We therefore assume that the failure modes of a node are restricted to crash failure semantics, i.e., the node is failsilent–it either operates correctly or not at all. We thus assume that the node implementation contains mechanism that ensure this fail-silent node behavior.

Event-Triggered Architectures

The communication protocols in ET-systems are event triggered. Since only the sender has knowledge about the point in time when a message has to be transmitted, the error detection is based on a time-out at the sender waiting for an acknowledgement message from the receiver (Positive Acknowledgement or Retransmission Protocol). From the receivers point of view, the states 'no activity by the sender' and 'failed sender' cannot be distinguished. In a pure ET-system a timely detection of node failures is difficult to achieve.

Time-Triggered Architectures

In a TT-system every node has to send a message at a predetermined point in time, otherwise it has failed. If the point in time, when a node has to send a message, is synchronized with the global time and known apriori to all receivers, then the error detection latency can be minimized.

3.5 Replica Determinism

Responsive systems are defined to be fault-tolerant, i.e., they have to provide the specified timely service despite the occurrence of faults specified in the fault-hypothesis. In many real-time applications the time needed to perform checkpointing and backward recovery after a fault has occurred is not available. Therefore fault-tolerance in responsive systems is normally based on active redundancy. Active redundancy requires replica determinism, i.e., the active replicas must take the same decisions at about the same time in order to maintain state synchronism. If replica determinism is maintained, fault-tolerance can be implemented by duplex fail-silent selfchecking nodes (or by Triple Modular Redundancy with voting if the fail-silent assumption is not supported).

Event-Triggered Architectures

Replica determinism is difficult to achieve in asynchronous ET-systems based on dynamic preemptive scheduling strategies. The slightly different processing speeds of two actively redundant nodes, caused by variations in the quartz crystals that drive the two autonomous CPUs, can bring about different preemption points in the two replicas in case a single external event requests immediate service. As a consequence, the state synchronism is lost.

Time-Triggered Architectures

The basic cycle time of a TT-system introduces a discrete time base of specified granularity. We call this discrete time base the action grid. Task changes are only permitted at the grid points of this action grid. They are either preplanned within a schedule or are the consequence of a global mode switch caused by a significant external event. Nondeterministic decisions can be avoided and replica determinism can be maintained without additional interreplica communication. TMR structures as well as selfchecking duplex nodes can be supported for the implementation of active redundancy without any difficulty.

3.6   Redundancy Management

If the required Mean Time To Failure (MTTF) of an application is substantially longer than the MTTF of the individual nodes, then the issues of repair and reintegration of repaired components have to be solved at the architectural level. In systems that have a high transient failure rate, the fast reintegration of failed components is of particular concern. Redundancy management is a critical activity in any fault-tolerant system. A failure in the redundancy management, e.g., the replacement of a correct component by a failed component, can have catastrophic consequences. In a responsive system the redundancy management has to be distributed. At the core of distributed redundancy management lies the solution of the distributed membership problem, i.e. the provision of a consistent view at all nodes about which node is present and which node is absent. Since the dissemination of the membership information takes some time, a careful distinction must be made between the point in time, about which a membership statement is being made (the membership point) and the point in time at which this statement can be made consistently at all nodes (the awareness point). A responsive membership protocol will have a short latency between the membership point and the awareness point. This latency is called the awareness latency.

Event-Triggered Architectures

The design of a responsive membership protocol in an asynchronous ET-architecture is a hard problem. Let us, for the moment, disregard the possibility of message loss. Then silence of a node in an ET-architecture can be caused either by the inactivity of the node (a correct state) or by a node failure (a failed state). The distinction between these two states is, however, crucial for the establishment of membership. Without the generation of additional periodic time triggered messages–we call them life-sign messages–the awareness latency in an ET architecture cannot be bounded.

Time-Triggered Architectures

In TT-architectures the periodic time-triggered messages can be regarded as life-sign messages for the purpose of membership establishment. The awareness latency can thus be bounded by the time it takes to complete a full TDMA round. The design of a responsive membership protocol is thus much easier in a TT architecture than in an ET architecture. Given an upto date membership status, it is clear how to accomplish the redundancy management. If the majority of nodes agree that a given node has failed (this is the result of the membership protocol) then this failed node can be replaced by a functioning node.

4　Conclusion

This paper identified six hard problems in the design of any responsive system: flow control, scheduling, testing for timeliness, timely error detection, replica determinism, and redundancy management. It then discussed possible solutions to these problems in event-triggered (ET) and time-triggered (TT) architectures. It is concluded that the solution to each one of these problems is easier in a TT architecture than an ET architecture. This is contrary to the prevailing opinion that states that the design of a real-time systems must be event triggered. It is hoped that this contribution will lead to a lively discussion and that a change in paradigm from event-triggered to time-triggered architectures will be stipulated if predictable timeliness and fault-tolerance are the stated goals in a safety critical application.

Acknowledgement

This work has been supported, in part, by ESPRIT Basic Research Project PDCS and by the Austrian Science Foundataion FWF project P8002 TEC.

References

[1] M. Malek, "Responsive systems: A challenge for the nineties, key-note address," *Proc. of EUROMICRO'90, 16th Symp. on Microprocessing and Microprogramming*, (Amsterdam, The Netherlands), pp. 9–16, August 1990.

[2] H. Kopetz and W. Ochsenreiter, "Clock synchronization in distributed real-time systems," *IEEE Transactions on Computers*, vol. 36, pp. 933–940, Aug. 1987.

[3] D. Powell, G. Bonn, D. Seaton, P. Verissimo, and F. Waeselynck, "The DELTA-4 approach to dependability in open distributed computing systems," *Proc. of 18th Int. Symposium on Fault-Tolerant Computing*, (Tokyo, Japan), pp. 246–151, June 1988.

[4] C. J. W. R. M. Kieckhafer, , and A. M. Finn, "Maft: A multicomputer architecture for fault-tolerance in real-time control systems," *Proc. of 5th Real-Time Systems Symposium*, pp. 133–140, Dec. 1984.

[5] H. Kopetz, A. Damm, C. Koza, M. Mulazzani, W. Schwabl, C. Senft, and R. Zainlinger, "Distributed fault-tolerant real-time systems: The MARS Approach," *IEEE Micro*, vol. 9, pp. 25–40, Feb. 1989.

[6] H. Kopetz and K. Kim, "Temporal uncertainties in interactions among real-time objects," *Proc. of 9th Symposium on Reliable Distributed Systems*, (Huntsville, AL, USA), pp. 165–174, IEEE Computer Society Press, Oct. 1990.

[7] F. Cristian, H. Aghili, R. Strong, and D. Dolev, "Atomic broadcast: From simple message diffusion to byzantine agreement," *Proc. of 15th Int. Symposium on Fault-Tolerant Computing*, (Silver Spring), pp. 200–206, June 1985.

[8] "PDCS, first year report," *University of Newcastle upon Tyne*, UK, May 1990. 3 Volumes.

[9] H. Kopetz and W. Merker, "The architecture of MARS," *Proc. of 15th Int. Symposium on Fault-Tolerant Computing*, (Ann Arbor, Michigan, USA), pp. 274–279, June 1985.

[10] A. K. Mok, *Fundamental Design Problems of Distributed Systems for the Hard Real-Time Environment*. PhD thesis, Massachusetts Institute of Technology, 1983. Report MIT/LCS/TR-297.

[11] C. L. Liu and J. W. Layland, "Scheduling algorithms for multiprogramming in a hard-real-time environment," *Journal of the ACM*, vol. 20, pp. 46–61, Jan. 1973.

[12] L. Sha, R. Rajkumar, and J. P. Lehoczky, "Priority inheritance protocols: An approach to real-time synchronization," *IEEE Transactions on Computers*, vol. C-39, pp. 1175–1185, September 1990.

[13] F. B. Schneider, "Implementing fault-tolerant services using the state machine approach: A tutorial," *ACM Computing Surveys*, vol. 22, pp. 299–319, Dec. 1990.

[14] J. Gebman, D. Mciver, and H. Shulman, "Maintenance data on the fire control radar," *Proc. of the 8th AIAA Avionics Conference*, (San Jose, USA), Oct. 1988.

[15] W. Schütz, "A test strategy for the distributed real-time system MARS," *IEEE CompEuro 90, Computer Systems and Software Engineering*, (Tel Aviv, Israel), pp. 20–27, May 1990.

Issues in Responsive Protocols Design

Y. KAKUDA   T. KIKUNO

Dept. of Information and Computer Sciences

Osaka University

Toyonaka, Osaka 560, Japan

{kakuda,kikuno}@ics.osaka-u.ac.jp

Abstract

Responsive protocols are communication protocols which ensure timely and reliable recovery when error events occur. In this paper, after a formal definition of responsive protocols is given, a conventional design method for real communication protocols is explained, and then design methods for responsive protocols using verification and synthesis are surveyed.

Key Words: Responsive System, Responsive Protocol, Protocol Engineering, Protocol Design, Protocol Verification, Protocol Synthesis

1 Introduction

Recently, production of communication protocols is a key issue that supports communication systems, and protocol engineering[8] has emerged. Since protocols include a large amount of abnormal processing triggered by faults, high reliability and performance in the presence of faults are required for such protocols. In order to satisfy this requirement, protocols must have fault-tolerant and real-time properties. According to the definition of responsive systems[9], protocols which satisfy these properties are called responsive protocols. This term was introduced in the First International Workshop on Responsive Computer Systems, which was held in 1991 at Golfe-Juan in France.

In this paper, first, responsive protocols which satisfy fault-tolerant and real-time properties are formally defined. Next, a conventional design method for responsive protocols is explained and as an example of a real protocol, a CCITT recommended protocol is shown. Finally, design methods for responsive protocols are surveyed. These are classified into two approaches: verification approach and synthesis approach. In the first approach, errors against fault-tolerant and real-time properties are detected and they are manually corrected, and after repetition of detection and correction, protocols are designed. In the second approach, protocols are designed by adding some functions into them such that fault-tolerant and real-time properties are satisfied. This paper discusses merits and demerits of both approaches.

2   Definition of Responsive Protocols

A protocol includes normal sequences of state transitions and abnormal sequences of state transitions. The normal sequences start from an initial state and end at normal terminal states. The abnormal sequences start from a normal nonterminal state where an error event occurs. If the protocol is fault-tolerant, then the abnormal sequences eventually revert to a normal nonterminal state. This phenomenon is called self-stabilization[2]. Otherwise, the abnormal sequences lead to abnormal terminal states or abnormal nonterminal states. If the protocol has fault-tolerant and real-time properties, then they efficiently return to a normal nonterminal state. These properties are called responsive.

The responsive protocols are characterized as follows: Normal states are global states reachable from an initial global state through a sequence of state transitions such that each state transition is not triggered by any error events. A sequence of state transitions from an initial global state to a normal state is said to be a normal sequence of state transitions. Abnormal states are defined as global states that are not normal states. A sequence of state transitions, such that the first state transition is triggered by an error event and all intermediate global states except the final global state are abnormal states, is called an abnormal sequence of state transitions. A protocol is said to be responsive if, for each abnormal sequence of state transitions in the protocol, the final global state in the sequence is a normal state and time required for executing all the state transitions in either the normal or abnormal sequence is less than or equal to a specified a priori, positive integer.

If there is an abnormal sequence in the protocol such that the final state is not a normal state, it is said that the protocol includes an error against fault tolerance. If there is a normal or abnormal sequence in the protocol such that time required for executing all the state transitions in the normal or abnormal sequence is greater than a specified a priori, positive integer, it is said that the protocol includes an error against real time. These errors are called errors against responsiveness.

3 Conventional Design Method for Responsive Protocols

Design of responsive protocols involves the following steps.

(1) Specification of protocol structure: A relation for message flow among processes is specified. For example, the protocol structure of a main part of the ISDN User Part (ISUP) of CCITT Signalling System No.7, which is a common channel signalling system between ISDN network exchanges, is shown in Figure 1. In the figure, circles denote processes, arrows represent channels, and labels attached to the arrows denote messages. This figure shows how a process CPCI interacts with a process CC by transmitting messages Setup_ind, Start_reset_ind and Release_conf and receiving messages Alert_req, Setup_resp and Release_req.

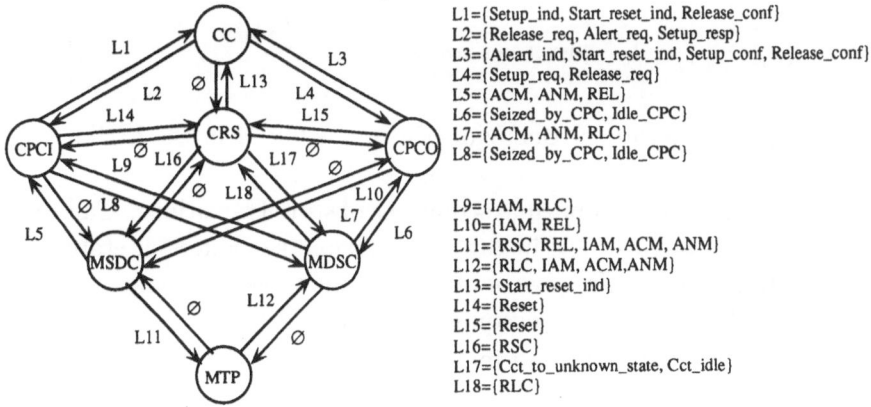

L1={Setup_ind, Start_reset_ind, Release_conf}
L2={Release_req, Alert_req, Setup_resp}
L3={Aleart_ind, Start_reset_ind, Setup_conf, Release_conf}
L4={Setup_req, Release_req}
L5={ACM, ANM, REL}
L6={Seized_by_CPC, Idle_CPC}
L7={ACM, ANM, RLC}
L8={Seized_by_CPC, Idle_CPC}

L9={IAM, RLC}
L10={IAM, REL}
L11={RSC, REL, IAM, ACM, ANM}
L12={RLC, IAM, ACM,ANM}
L13={Start_reset_ind}
L14={Reset}
L15={Reset}
L16={RSC}
L17={Cct_to_unknown_state, Cct_idle}
L18={RLC}

Figure 1: Protocol Structure of a Main Part of the ISUP.

(2) Specification of protocol behavior: A relation between transmission and reception of messages among processes is specified. For example, the protocol behavior of process CPCI is shown in Figure 2. In

the figure, circles denote states, arrows correspond to state transitions, and a minus sign denotes transmission of a message, and a plus sign indicates reception of a message. This figure illustrates that a process CPCI receives a message IAM at state IDLE and enters state S1, and transmits a message Seized_by_CPC at state S1 and enters state S2. In other words, the specification of protocol behavior is equivalent to the specification of sequences of state transitions. This step is divided into the following substeps.

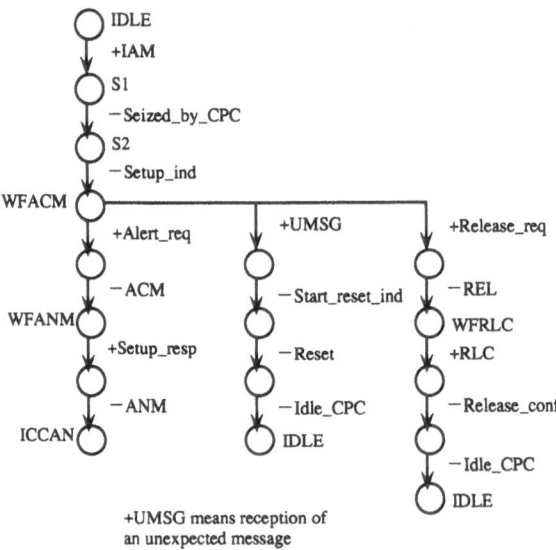

Figure 2: Protocol Behavior of Process CPCI in the ISUP.

(2-1) Specification of normal sequences of state transitions. Call setup and clearing sequences are instances of normal sequences of state transitions. In Figure 2, a sequence of state transitions from state IDLE to state ICCAN is a call setup sequence of the ISUP.

(2-2) Specification of abnormal sequences of state transitions. Reset sequences due to reception of unexpected messages are instances of abnormal sequences of state transitions. In Figure 2, a sequence of state transitions from state WFACM to state IDLE through +UMSG is a reset sequence of the ISUP when process CPCI receives an unexpected message from process CC. Figure 3 shows a sequence chart representing the reset sequence triggered by process CPCI.

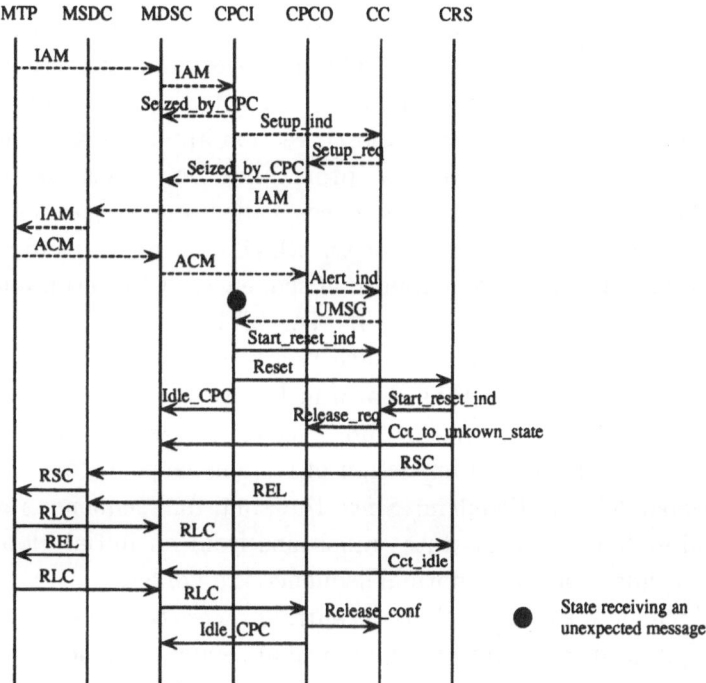

Figure 3: Sequence Chart of Reset Sequence Triggerd by Process CPCI.

It is important to observe that the number of normal sequences is restricted to the number which protocol designers can perfectly specify, while there are many abnormal sequences because error events can occur at any normal states. It is therefore important to specify abnormal sequences for design of responsive protocols.

As the step (1) and the substep (2-1) are common to design of any protocols, responsive or not, this paper focuses on the substep (2-2), which is crucial in the design of responsive protocols.

The substep (2-2) includes a) detection of errors against responsiveness in the specification of abnormal sequences of state transitions, and b) correction of the errors in it. The detection and correction are repeated until the errors are completely corrected. In the conventional design method, these detection and correction are made manually.

When error events occur, real protocols often revert to an initial global state through abnormal sequences, called reset sequences, of state transitions. There is a central process, which controls transmission and reception of messages in the abnormal sequences. Call setup and clearing protocols in the ISUP and the INMARSAT satellite communication systems are typical examples of such real protocols. Process CRS is a central process of the ISUP as shown in Figure 1.

In the above conventional methods using reset sequences, there are two problems to be solved for design of responsive protocols.

(1) Dependability Problem: Since the abnormal sequence reverts to an initial global state, all state transitions from an initial global state to an abnormal state in a normal sequence are voided.

(2) Timeliness Problem: Since a central process controls transmission and reception of messages in the abnormal sequence, time required for executing all the state transitions in the abnormal sequence may exceed an assumed deadline.

4   Verification Method for Responsive Protocols

One approach to solve the Dependability and Timeliness Problems is a design method for responsive protocols using verification techniques. These techniques aim to prove that the specification meets requirements for responsive protocols. In this approach, after the specification has been designed without any restrictions, errors against responsiveness are corrected manually. The specification is flexibly designed by placing no restrictions on it. However, the verification of a large and complicated specification takes a large amount of time or an infinite time in the worst case.

The typical work on verification of responsive protocols was carried out by Gouda and Multari[3]. They identify some important characteristics of self-stabilized protocols and propose a mathematical method for verifying self-stabilization of protocols. In addition, they show how to design a number of self-stabilized protocols using closure and convergence properties[1]. However, they did not discuss the real-time properties for design of responsive protocols. Kakuda and Kikuno propose an automated method for verifying responsive protocols[4]. The protocols are modeled by an extended finite state machine, which allows to use variables, but values of the variables are restricted to finite. In order to verify responsive protocols, they apply methods for protocol verification to verification of fault-tolerant properties, and task scheduling algorithms for multi-processor systems to verification of real-time properties. Kawashima, Kakuda and Kikuno improve the automated method using real-time convergence property for verifying responsive protocols with infinite values of variables[7]. Saito and Hasegawa discuss applicability of the protocol validation tool developed by KDD to responsive protocols[10]. The usefulness of this tool has been proved by showing design examples of real protocol specifications using this tool. It is thus expected that an extention of this tool will support design of real responsive protocols.

All the design methods for responsive protocols described above are application specific. They are flexible but have difficulties in requiring expertise in protocols to be corrected, encompassing all failure modes, and establishing time bounds for recovery. Furthermore, they take a large amount of time for verification of a large and complicated protocols.

5   Synthesis Method for Responsive Protocols

The other approach to solve the Dependability and Timeliness Problems is a design method for responsive protocols using synthesis techniques. The aim of these techniques is to add functions for responsive protocols to the specification. In this approach, the specfication can be designed efficiently at the expense of the flexibility. The designed specfication is however not always the one that protocol designer has in mind. In that case, the designer must correct the specification.

Methods for synthesis of responsive protocols were studied by two research groups. Saleh, Ural and Agarwal discuss applicability of the distributed snapshots algorithm to self-stabilization of protocols[11] and suggest a modified distributed snapshots algorithm for stabilizing protocols. Kakuda, Kikuno, Malek and Saito propose a method for synthesis of responsive protocols using checkpointing and rollback-recovery[5]. In this method, sequences of state transitions obtained by the rollback-recovery form abnormal sequences of state transitions in the responsive protocol.

The proposed method using checkpointing and recovery[5] is suitable for design of responsive protocols for the following reasons.

(1) Fault-Tolerance: Since any abnormal sequence in the protocol can revert to an intermediate normal state while retaining consistency on transmission and reception of messages, only state transitions from an intermediate normal state to an abnormal state in a normal sequence are voided.

(2) Timeliness: Since any process can initiate rollback-recovery and any abnormal sequence in the protocol can efficiently revert to the intermediate normal state using the broadcast mechanism, time required for executing all state transitions in the abnormal sequence is shorter and will more likely meet a desired deadline. The designed protocols have a fixed time bound on recovery and the recovery time is optimized[6].

All the design methods for responsive protocols described above are application independent. Since they only require the protocol structure and exchange of local state information, they are regarded as unified design methods for any protocols and all failure modes. However, they assume that fault-tolerance of the procedure and memory required for the synthesis methods are insured.

6   Conclusion

This paper has defined responsive protocols and surveyed methods for design of responsive protocols using verification and synthesis. The former methods tend to be flexible and application specific and the latter methods tend to be efficient and application independent. However, all the methods have merits and demerits. Advanced design methods for responsive protocols which have merits in both previous methods should be extensively studied.

References

[1] A. Arora and M. Gouda, "Closure and convergence: a foundation of fault-tolerant computing," *Proc. of Int'l. Symp. on Fault-Tolerant Computing*, pp.396-403, June 1992.

[2] E. W. Dijkstra, "Self-stabilizing systems in spite of distributed control," *Communications of the ACM*, 17, 11, pp.643-644, Nov. 1974.

[3] M. G. Gouda and N. J. Multari, "Stabilizing communication protocols," *IEEE Trans. on Computers*, 40, 4, pp.448-458, April 1991.

[4] Y. Kakuda and T. Kikuno, "Verification of responsiveness for communication protocols," *IEICE Japan, Tech. Group Paper*, CPSY 91-58 (FTS 91-57), Dec. 1991.

[5] Y. Kakuda, T. Kikuno, M. Malek and H. Saito, "A unified approach to design of responsive protocols," *Proc. the 1992 IEEE Workshop on Fault-Tolerant Parallel and Distributed Systems*, pp.8-15, Amherst, Massachusetts, July 1992.

[6] Y. Kakuda, T. Kikuno, M. Malek and H. Saito, "Efficient checkpointing for protocol recovery in communication systems," *Proc. the 1991 International Symposium on Communications* (ISCOM'91), pp.704-707, Tainan, Taiwan, Dec. 1991.

[7] K. Kawashima, Y. Kakuda and T. Kikuno, "Verification of responsive protocols based on real-time convergence of sequences of predicates," *IEICE Japan, Tech. Group Paper*, CPSY 91-82, March 1992, in Japanese.

[8] M. T. Liu, "Protocol engineering," *Advances in Computers*, vol. 29, pp.79-195, 1989.

[9] M. Malek, "Responsive systems: A challenge for the nineties," *Proc. EUROMICRO'90, 16th Symp. on Microprocessing and Micropro-gramming,* Keynote Address, Amsterdam, The Netherlands, North-Holland, Microprocessing and Microprogramming 30, pp.9-16, Aug. 1990.

[10] H. Saito and T. Hasegawa, "Protocol validation tool and its appli-cability to responsive protocols," *Proc. of the Second International Workshop on Responsive Computer Systems,* Oct. 1992, Springer-Verlag, Dependable Computing and Fault-Tolerant Systems, 7, pp.207-222.

[11] K. Saleh, H. Ural, and A. Agarwal, "A distributed snapshots algo-rithm and its application to protocol stabilization," *Proc. of the Sec-ond International Workshop on Responsive Computer Systems,* Oct. 1992, Springer-Verlag, Dependable Computing and Fault-Tolerant Systems, 7, pp.197-206.

Responsive Systems Theory

A Probabilistic Duration Calculus*

Z. LIU[a] A. P. RAVN[b] E. V. SØRENSEN[b] C. ZHOU[c,d]

a Department of Computer Science,University of Warwick
Coventry CV4 7AL, England, UK
b Department of Computer Science,Technical University of Denmark
DK-2800 , Lyngby, Denmark
c International Institute for Software Technology,United Nations University
P.O.Box 3058, Macau
d On leave from Software Institute, Academia Sinica
Beijing, P.R. China

Abstract

This paper presents a calculus that enables a designer of an embedded, real-time system to reason about and calculate whether a given requirement will hold with a sufficient high probability for given failure probabilities for components used in the design of the system. The main idea is to specify requirements and design in Duration Calculus, a real-time, interval logic, to define satisfaction probabilities for formulas in this calculus, and establish a calculus with rules that support calculation of the probability for a composite formula from probabilities of its constituents. This ensures that reasoning about probabilities is consistent with requirements and design decisions. We thus avoid introducing separate models for requirements and reliability analysis. The system model is a finite automaton with fixed transition probabilities. This defines discrete Markov processes as basis for the calculus.

Key Words: duration calculus, real-time systems, probabilistic automata, satisfaction probability.

---
*This research was supported in part by **ProCoS** ESPRIT BRA 3104, by the Danish Technical Research Council under project **RapID**, and by the research grants GR/D11521 and GR/H39499 from the Science and Engineering Research Council of UK.

1   Introduction

Requirements for an embedded, real-time system include functional and safety properties. Consider for instance an on-off gas burner [1]. It is required to turn the flame on or off a short time after requested to do so by a thermostat. It must also prevent excessive leak of gas to the environment. The latter requirement can be stated as an integrated constraint: the duration of leaking states should only be a small proportion of any 1 minute interval.

Such a system can be modelled by a dynamic system where a state changes over time. In the gas burner example, we could for instance introduce a state *leak*, that varies with time. A design for discrete control of the system will then be given by constraints on the transitions between states. A designer may now use various mathematical techniques to verify that the design satisfies the requirements. Among them the duration calculus [2] is recently found promising for reasoning about requirements and designs of real-time, embedded systems [3, 4, 5]. A summary of this calculus is given in Section 2.

However, a customer or a certification agency may legitimately ask about the dependability of the system in terms of a failure probability within a certain period of time. Such a question cannot be answered from the design or its mathematical model. In order to answer the questions, the designer may choose to develop alternative models, cf. the two tiered approaches used in the SIFT project [6], or the stochastic model developed from a state machine model for a design in [1].

A two model approach adds complexity to the design activity because the two have somehow to be updated consistently whenever the design changes. Several researchers have seen that there is a potential for making the design activity simpler by using a unified model in the form of a probabilistic automaton with Markov properties [7, 8]. In [8] an untimed logic for specification is extended by adding probabilities to the combinators; this allows reasoning about untimed probabilistic systems. Time and probabilities are introduced together in [7] which extends the computation tree logic (CTL) of [9]. There is however not a proof system for the extended logic, and the expressiveness is somewhat restricted. We have thus found it worthwhile to investigate the development of a probabilistic duration calculus.

Based on probabilistic automata, this paper defines the satisfaction probabilities of duration formulas in the duration calculus, and establishes a corresponding probabilistic calculus. The calculus has a set of axioms and rules that support direct reasoning about and calculation of satisfaction probabilities for formulas specifying a given design.

The probabilistic duration calculus is based on three key ideas. The first is to simulate imperfect systems with probabilistic automata. This is presented in Section 3. The second one, in Section 4, is to extend the model of the duration calculus and define the satisfaction probability of a duration formula by a probabilistic automaton. And the third one is to establish a calculus to calculate and reason about satisfaction probabilities. This is done in section 5. Running examples are given in each section and Section 6 contains a number of examples that illustrate the possible application of the calculus. The conclusion in Section 7 compares this work with related work.

2 The Duration Calculus

This section outlines the duration calculus and its application to specifying real-time systems.

2.1 Time

The original duration calculus [2, 10] uses continuous time. In order to have a simple, well understood probabilistic model (see Section 3), we here assume discrete time. Time is represented by the set N of non-negative integers. A time point is denoted t, t_1, etc. and a *time interval* $[t_1, t_2]$, $t_1 \leq t_2$, represents the set of time points from t_1 to t_2.

2.2 States

We assume a finite non-empty set A of *primitive states* which are boolean variables, defined for every time point. States, ranged over by P, Q, P_1, Q_1, etc., consist of expressions formed by the following rules:

- Each primitive state $P \in A$ is a state.

- If P and Q are states, then so are $\neg P$, $(P \wedge Q)$, $(P \vee Q)$, $(P \Rightarrow Q)$, $(P \Leftrightarrow Q)$.

A primitive state P is interpreted by an *interpretation I* as a function $I(P) : N \to \{0,1\}$. $I(P)(t) = 1$ means that state P is present at time point t, and $I(P)(t) = 0$ means that state P is not present at time point t. We assume that when a state is present at time t, it will persist for the next time unit. A *composite state* is interpreted as a function which is defined by the interpretations for the primitive states and the boolean operators.

2.3 Duration

For an arbitrary state P, its *duration* is denoted $\int P$. Given an interpretation I of the states, a duration $\int P$ will be interpreted over time intervals. It denotes the accumulated time P is present within the time interval. So for an arbitrary interval $[t_1, t_2]$, the interpretation $I(\int P)([t_1, t_2])$ is defined as the non-negative integer

$$I(\int P)([t_1, t_2]) = \sum_{t=t_1}^{t_2-1} I(P)(t)$$

where $I(\int P)([t, t]) = 0$. So $\int 1$ always denotes the length of an interval.

The set of *primitive duration terms* consists of variables over the integers Z and durations of states. A *duration term* is either a primitive term or an expression formed from terms by using the usual operators on integers, such as addition $+$ and multiplication $*$.

2.4 Duration Formulas

A *primitive duration formula* is an expression formed from terms by using the usual relational operators on the integers, such as equality $=$ and inequality $<$. A *duration formula* is either a primitive formula or an expression formed from formulas by using the logical operators $\neg$, $\wedge$, $\vee$, $\Rightarrow$, $\Leftrightarrow$, and the *chop* ; and quantifiers $\forall$, $\exists$ applied to variables ranging over Z.

A duration formula D is satisfied by an interpretation I with an interval $[t_1, t_2]$ just when it is evaluated to true [10]. There it is written

$$I, [t_1, t_2] \models D$$

A chopped formula $D_1; D_2$ is true for I with $[t_1, t_2]$ if there exists a t such that $t_1 \leq t \leq t_2$ and D_1 and D_2 are true respectively with $[t_1, t]$ and $[t, t_2]$ for I.

We define shorthands for some duration formulas which are often used.

Definition 1 *For an arbitrary state P,*

$$\lceil P \rceil \triangleq (\int P = \int 1) \wedge (\int 1 > 0)$$

This means that P holds everywhere in a non-point interval. We use $\lceil \ \rceil$ to denote the predicate which is true only for a point interval.

Definition 2 $\lceil \ \rceil \triangleq \int 1 = 0$

Definition 3 *For a duration formula D,*

$$\Diamond D \triangleq tt; D; tt$$

where tt is the duration formula which is true of any interval under any interpretation.

This is true of an interval in which D holds for some subinterval of it.

Definition 4 *For a duration formula D,*

$$\Box D \triangleq \neg \Diamond \neg D$$

This is true of an interval where D holds for all subintervals of it.

2.5 Real-Time Specifications

The duration calculus has been used to specify real-time constraints of embedded systems [3, 4, 5]. In [2], one of the time critical requirements of a Gas Burner is specified by a formula of the duration calculus denoted as **Req-1**,

$$\textbf{Req-1} \quad \int 1 > 60sec \Rightarrow (20 * \int leak \leq \int 1)$$

This says that if the interval over which the system is observed is at least one minute, the proportion of time spent in the leak state is no more than one twentieth of the elapsed time.

The requirement is refined into two design decisions

$$\textbf{Ds-1} \quad \Box (\lceil leak \rceil \Rightarrow \int 1 \leq 1sec)$$
$$\textbf{Ds-2} \quad \Box (\lceil leak \rceil; \lceil \neg leak \rceil; \lceil leak \rceil \Rightarrow \int 1 > 32sec)$$

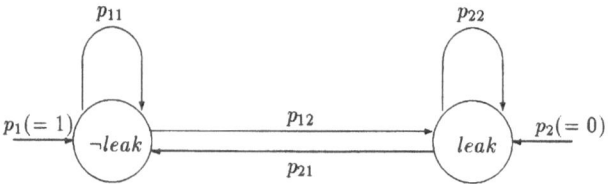

Figure 1: A Gas Burner With Unreliable Detector

Ds-1 says that any leak state must be detected and stopped within one second, and **Ds-2** says leak must be seperated by at least 30 seconds.

The correctness of the design is reasoned about by proving the implication

$$\textbf{Ds-1} \wedge \textbf{Ds-2} \Rightarrow \textbf{Req-1}$$

in the duration calculus [2].

However, we cannot expect, in practice, a *real* implementation to satisfy the decisions at all time. A real implementation can only satisfy the design decisions with some probability within a given service period. This raises the following problems which are the concerns of this paper. How can we model a real (imperfect) implementation? How can we define and reason about the satisfaction probability of a duration formula (requirement or decision)?

3   Imperfect Systems and Probabilistic Automata

We will use a *finite probabilistic automaton* as a mathematical model of the behaviour of an imperfect system in a discrete time domain. Such an automaton is well described by its transition graph. We will continue with the Gas Burner example.

3.1   Unreliable Flame Detector

A model of a Gas Burner with an unreliable flame detector can be defined by the transition graph shown in Figure 1. In this Gas Burner, *leak* is the only primitive state considered, and ¬*leak* denotes the absence of the primitive state. The probabilities of the system starting in states ¬*leak* and *leak* are p_1 and p_2 respectively[1], where $0 \leq p_1, p_2 \leq 1$ and $p_1 + p_2 = 1$. The probability of the system to stay burning within one time unit is p_{11}. The probability of flame failure within one time unit is p_{12}. So $0 \leq p_{11}, p_{12} \leq 1$ and $p_{11} + p_{12} = 1$. The probability of the detector to detect the leakage (thereby causing re-ignition the flame) within one time unit is p_{21}. The probability with which the detector fails to detect the leakage within one time unit is p_{22}, where $0 \leq p_{21}, p_{22} \leq 1$ and $p_{21} + p_{22} = 1$. Here we assume that the transition probabilities are independent of the transition history. This is the main feature of a Markov chain.

3.2   Unreliable Detector and Ignition

An implementation with more imperfect components is modelled by a larger graph. The model of a Gas Burner with an unreliable flame detector and an unreliable ignition can be illustrated in Figure 2. This graph[2] uses two primitive states *gas* (the gas and ignition is switched on), and *flame* (the flame is on) to model the system:

- It starts in the idle state, i.e. both the gas and the flame are off,

$$p_1 = 1 \quad \text{and} \quad p_2 = p_3 = p_4 = 0$$

- It idles with probability p_{11} for one time unit;

- The ignition succeeds with probability p_{12} within one time unit;

- The ignition fails with probability p_{13} within one time unit;

- The system stays burning with probability p_{22} within one time unit;

- The system finishes service with probability p_{21} within one time unit;

---

[1]We usually assume that the Gas Burner starts from the state ¬*leak*, i.e. $p_1 = 1$ and $p_2 = 0$.

[2]Here and in Section 3.1, we are using *complete graphs* in the figures to illustrate the model. For simplicity, we can eliminate *unreachable* states such as the one labelled by ¬*gas* ∧ *flame* in Figure 2, to which the initial probability and the probabilities of transitions to it are zero, and eliminate all the impossible transitions which depart from a unreachable state or have probabilities zero (see Section 6.2).

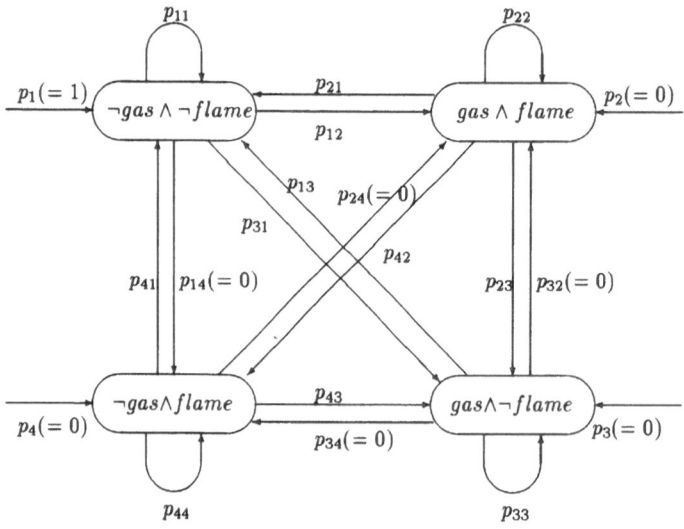

Figure 2: A Gas Burner With Unreliable Detector and Ignition

- The system detects and stops a failure (by returning to the idle state) with probability p_{31} within one time unit;

- The flame fails with probability p_{23} within one time unit;

- Detector or recovery fails with probability p_{33} within one time unit.

We have $0 \leq p_{ij} \leq 1$ and

$$p_{i1} + p_{i2} + p_{i3} + p_{i4} = 1 \quad (i = 1, 2, 3, 4)$$

leak is now the composite state

$$leak \triangleq gas \wedge \neg flame$$

3.3   Probabilistic Automaton

We end this section with a general definition of a probabilistic automaton (PA).

Definition 5 *A PA is a tuple* $G = (A, \tau_0, \tau)$ *where*

- *A is a finite non-empty set of primitive states.*

  A basic conjunction[3] of A defines a state of the system, which deter-mines the presence and absence of the primitive states. The set of all basic conjunctions is denoted by V, ranged over by v, v', v_i, etc.

- *τ_0 : V → [0, 1] is a function called the initial probability mass function, such that*

$$\sum_{v \in V} \tau_0(v) = 1$$

  $\tau_0(v)$ defines the probability of the system starting from state v.

- *τ : V × V → [0, 1] is a function called the single-step probabilistic transition function such that for every v ∈ V*

$$\sum_{v' \in V} \tau(v, v') = 1$$

For example, in Section 3.1, $A = \{leak\}$, $V = \{leak, \neg leak\}$. The initial probability mass function is $\tau_0(\neg leak) = 1$, $\tau_0(leak) = 0$. And the probabilistic transition function is as follows.

$$\tau(\neg leak, \neg leak) = p_{11} \quad \tau(\neg leak, leak) = p_{12}$$
$$\tau(leak, \neg leak) = p_{21} \quad \tau(leak, leak) = p_{22}$$

4 Satisfaction Probability

Given a $G = (A, \tau_0, \tau)$ where V is the set of all basic conjunctions of A.

4.1 Probabilistic Behaviour

An infinite sequence of states in V,

$$\sigma : v_1, \ldots, v_n, \ldots,$$

defines a possible *behaviour* of G. That is, start from state v_1 and transit from state v_{i-1} to v_i within one time unit. We will use $\sigma(i)$, $i \geq 1$, to denote the ith state v_i in σ. Let X denote the set of all behaviours of G, ranged over by σ, γ, etc. We use $\sigma^{[n]}$ to denote the prefix of length

---

[3]A conjunction of A is *basic*, if every primitive state in A or its negation, but not both, appears in the conjunction.

n of σ, and $X(\sigma^{[n]})$ to denote the set of behaviours with common prefix $\sigma^{[n]}$. Note that $X = X(\sigma^{[0]})$ for any σ.

τ_0 determines the probability of the set of behaviours starting from an initial state, and τ determines the probability of a transition from one state to another. Therefore, τ_0 and τ together determine the probability of the set $X(\sigma^{[n]})$ of behaviours with the common prefix $\sigma^{[n]}$ of length n. We denote its probability by $\mu(X(\sigma^{[n]}))$. Thus

$$\mu(X(\sigma^{[n]})) \triangleq \tau_0(\sigma(1)) * \prod_{i=1}^{n-1} \tau(\sigma(i), \sigma(i+1))$$

where $\mu(X(\sigma^{[0]})) \triangleq 1$ and $\mu(X(\sigma^{[1]})) \triangleq \tau_0(\sigma(1))$

For example, let $\sigma^{[1]} = leak$ in Section 3.1. $X(\sigma^{[1]})$ is then the behaviours starting from state $leak$. But $\tau_0(leak) = 0$, i.e. the system cannot start from state $leak$. Thus, $\mu(X(\sigma^{[1]})) = 0$. Similarly, let

$$\sigma^{[5]} : \neg leak, \neg leak, leak, leak, \neg leak$$

be another prefix. The probability of behaviours with $\sigma^{[5]}$ as common prefix is

$$\mu(X(\sigma^{[5]})) = p_1 * p_{11} * p_{12} * p_{22} * p_{21}$$

We now define a probability space $\langle X, \mathcal{A}, \mu \rangle$, where $\mathcal{A}$ is the sigma-algebra[4] generated by the sets $X(\sigma^{[n]})$. The definition of μ on the generators[5] $X(\sigma^{[n]})$ given above uniquely determines the *probabilistic measure* [11] which is the extension of μ (also denoted μ) to all sets in the sigma-algebra $\mathcal{A}$. This definition of the probability space makes the probabilistic automaton a discrete Markov chain [11].

4.2   Satisfaction

A behaviour σ of G determines presence and absence of the primitive states at each time point, and thus defines an interpretation I_σ of dura-

---

[4]That is, $\mathcal{A}$ is a family of subsets of X such that: (i) If $X_1 \in \mathcal{A}$ so is its complement $X - X_1$. (ii) If $\{X_n\}$ is any countable collection of sets in $\mathcal{A}$, then also their union $\bigcup X_n$ and their intersection $\bigcap X_n$ belong to $\mathcal{A}$ (see [11] for details.

[5]$\mathcal{A}$ is generated by the sets $X(\sigma^{[n]})$ means that $\mathcal{A}$ is the smallest (*w.r.t.* the inclusion relation between families of subsets) sigma-algebra containing all the sets $X(\sigma^{[n]})$, for $n \geq 0$. Each $X(\sigma^{[n]})$ is a generator of $\mathcal{A}$.

tion formulas with A as the primitive states. That is,

$$I_\sigma(P)(j) \triangleq \begin{cases} 1 & \text{if } \sigma(j) \Rightarrow P \\ 0 & \text{if } \sigma(j) \Rightarrow \neg P \end{cases}$$

where $j \in N$.

This gives the following definition of satisfaction of a duration formula D (with A as primitive states) by a behaviour σ of G with a time interval $[t_1, t_2]$,

$$\sigma, [t_1, t_2] \models D \quad \text{iff} \quad I_\sigma, [t_1, t_2] \models D$$

For example, let $\sigma^{[5]}$ be

$$\neg leak, \neg leak, leak, leak, \neg leak$$

Then for every behaviour $\gamma \in X(\sigma^{[5]})$ with $\sigma^{[5]}$ as its prefix, we have for the time interval $[0, 5]$, γ satisfies $\int 1 = 5$, i.e. γ spends 5 units of time in this interval

$$\gamma, [0, 5] \models \int 1 = 5$$

For the interval $[0, 5]$, γ satisfies that the total time spent in the $\neg leak$ state is 3 units

$$\gamma, [0, 5] \models \int \neg leak = 3$$

During any sub-interval of $[0, 5]$, γ may never spend more than 2 units of time in $leak$ state

$$\gamma, [0, 5] \models \Box(\lceil leak \rceil \Rightarrow \int 1 \leq 2)$$

For $[0, 5]$, γ spends 5 units but not 3 units of time

$$\gamma, [0, 5] \not\models \int 1 \leq 3$$

Since $leak$ holds for time points 2 and 3, γ satisfies $\Diamond \lceil leak \rceil \wedge (\int 1 = 2)$ for the interval $[0, 5]$ and thus

$$\gamma, [0, 5] \not\models \Box(\lceil leak \rceil \Rightarrow \int 1 \leq 1)$$

4.3   Satisfaction Probability

The probability of a PA satisfying a requirement (a duration formula) within a certain operating time (starting from time zero) is defined as the probability of the set of behaviours of the system which satisfy the requirement within this interval. Let D be a duration formula, and $[0, t]$ be a prefix interval of N. Then $\mu(D)[0, t]$, denoting the *satisfaction probability* of D by G within the time interval $[0, t]$, is defined

$$\mu(D)[0, t] \triangleq \mu(\{\sigma \in X \mid \sigma, [0, t] \models D\})$$

Note that whether a behaviour σ satisfies D within $[0, t]$ only depends upon its prefix $\sigma^{[t]}$ of length t. Therefore, the set on the right hand side of the above definition is the union of a finite number of generators, say $X(\sigma_1^{[t]}), \ldots, X(\sigma_m^{[t]})$, of the sigma-algebra $\mathcal{A}$ in the probability space $\langle X, \mathcal{A}, \mu \rangle$, or it is an empty set. That is, it is a member of $\mathcal{A}$.

For example, for the PA defined in Section 3.1, let $D \triangleq \square(\lceil leak \rceil \Rightarrow \int 1 \leq 1)$. Then within $[0, 2]$, the behaviours satisfying D are the union of the sets

$$X(\neg leak, \neg leak), \quad X(\neg leak, leak), \quad X(leak, \neg leak)$$

Thus

$$\begin{aligned} \mu(D)[0, 2] &= p_1 * p_{11} + p_1 * p_{12} + p_2 * p_{21} \\ &= p_{11} + p_{12} \\ &= 1 \end{aligned}$$

5   PDC: Probabilistic Duration Calculus

This section establishes a calculus about the satisfaction probability $\mu(D)$, and the length T of a prefix interval over N.

The *probabilistic duration logic* is based on a first order modal logic [12] and real arithmetic with additional interval functions T and $\mu(D)$s. Let Int be the set of the finite prefix intervals over N. The function T belongs to $Int \rightarrow N$ and assigns each prefix interval $[0, t]$ with its length t, while $\mu(D)$ belongs to $Int \rightarrow [0, 1]$ which assigns each prefix interval $[0, t]$ with the satisfaction probability of duration formula D within $[0, t]$.

Therefore, in this logic a *primitive term* is T, $\mu(D)$, or a variable x ranging in the real numbers. A *term* is a primitive term, or an expression of terms built using the usual operators on real numbers, such as addition $+$ and multiplication $*$, with their standard meanings.

A *primitive formula* is an expression built from terms using the relational operators, such as equal $=$ and less than $<$ with their standard meanings.

A *formula* is a primitive formula or an expression built from formulas using the first order logic operators, a modal operator $\Diamond_p$ and the quantifiers over variables.

The modal operator $\Diamond_p$ is interpreted as: $\Diamond_p F$ holds for an interval $[0, t]$ iff there is a prefix $[0, t_1]$, $t_1 \leq t$, such that F holds for $[0, t_1]$. We assume the standard interpretations for the rest of the logic operators and quantifiers.

In this logic, we can write down and reason about probabilistic formulas such as

$$\mu(\neg\textbf{Req-1}) \leq \mu(\neg\textbf{Ds-1}) + \mu(\neg\textbf{Ds-2})$$

which asserts that the probability of violating the requirement will not be greater than the sum of the probabilities of violating the design decisions. This formula tells the designer that there is a trade off between the design decisions with respect to probabilities. It also allows the designer to consider the reliability of each one separately.

Satisfaction probabilities can also be calculated with this logic by reasoning about formulas of the form

$$T = t \Rightarrow \mu(D) = p$$

As an extension, PDC will include all axioms and rules from the modal logic and real arithmetic. We present in what follows the additional ones for T and $\mu(D)$.

The duration formula tt defines the set X of all behaviours of G for any interval.

AR 1 *For the duration formula tt*

$$\mu(tt) = 1$$

For any given interval, the sets of behaviours defined by D and $\neg D$ form a partition of all the behaviours X. So the sum of their probabilities is 1.

AR 2 *For an arbitrary duration formula D*

$$\mu(D) + \mu(\neg D) = 1$$

The following axiom formalizes the additivity rule in probability theory.

AR 3 *For arbitrary duration formulas D_1 and D_2*

$$\mu(D_1 \vee D_2) + \mu(D_1 \wedge D_2) = \mu(D_1) + \mu(D_2)$$

The satisfaction probability is monotonic in the sense that

AR 4 *If $D_1 \Rightarrow D_2$ holds in the duration calculus, then $\mu(D_1) \leq \mu(D_2)$ holds in PDC.*

The above four axioms and rules follow directly from probability theory. The following theorem can easily be proven from them.

Theorem 1 *For arbitrary duration formulas D, D_1, D_2 and D_3*

1. $\mu(\int f) = 0$

2. $0 \leq \mu(D) \leq 1$

3. *If $D_1 \Leftrightarrow D_2$ in the duration calculus, then $\mu(D_1) = \mu(D_2)$*

4. *If $D_1 \wedge D_2 \Rightarrow D_3$ in the duration calculus, then*

$$(\mu(D_1) = 1) \Rightarrow (\mu(D_2) \leq \mu(D_3))$$

No behaviour with a prefix of length t satisfies $\int 1 \neq t$.

AR 5 $T = t \Rightarrow (\mu(\int 1 \neq t) = 0)$

The validity of a *global formula* (no modal terms T or $\mu(D)$ in it) does not depend upon the time interval. More generally we have

AR 6 *For a PDC formula F such that t is not free in F or F is a global formula, if*

$$(T = t) \Rightarrow F$$

holds, then F holds.

Within $[0, t]$, D and $(D \wedge \int 1 = t); tt$ are satisfied by the same behaviours.

AR 7 *For an arbitrary duration formula D*

$$T = t \Rightarrow (\mu(D) = s) \quad iff \quad T \geq t \Rightarrow \mu(D \wedge (\int 1 = t); tt) = s$$

$\Diamond_p F$ means a prefix segment satisfying F. So we have the rule:

AR 8 *For an arbitrary PDC formula F without u as a free variable,*

$$\exists u \leq t : T = u \Rightarrow F \ \ iff \ \ T \geq t \Rightarrow \Diamond_p F$$

We now can prove the following theorem.

Theorem 2 *For arbitrary duration formulas D, D_1 and D_2*

1. $T = t \Rightarrow (\mu(D) = \mu(D \wedge \int 1 = t))$

2. $T = t \Rightarrow (\mu(\int 1 = t) = 1)$

3. *If $\mu(D_1) = 0$, then $\mu(D_1; D_2) = 0$*

The axioms and rules described so far are independent of the Markov properties of the PA defined by the probability space $\langle X, \mathcal{A}, \mu \rangle$. We consider, in this paper, only those PAs which are Markov chains. The two following axioms formalize the Markov properties for a PA $G = (A, \tau_0, \tau)$.

AR 9 *For an arbitrary state $v \in V$,*

$$T = 1 \Rightarrow (\mu(\lceil v \rceil^1) = \tau_0(v))$$

Here we have used the convention

$$\lceil v \rceil^1 \triangleq \lceil v \rceil \wedge (\int 1 = 1)$$

This axiom formalizes the initial probability mass function τ_0. The probabilistic transition function τ is formalized as follows.

AR 10 *For an arbitrary duration formula D and states $v_i, v_j \in V$,*

$$\mu(D; \lceil v_i \rceil^1; \lceil v_j \rceil^1) = \tau(v_i, v_j) * \mu(D; \lceil v_i \rceil^1; \int 1 = 1)$$

This provides a way for calculating the probability of behaviours by chopping of unit intervals.

As mentioned before, for a given interval $[0, t]$, the set of behaviours satisfying a duration formula D can be partitioned into the union of a finite number of generators of the sigma-algebra $\mathcal{A}$. Each generator $X(\sigma_i^{[t]})$ can be defined by a duration formula

$$\lceil v_{i1} \rceil^1; \ldots; \lceil v_{it} \rceil^1$$

where $v_{ij} = \sigma_i^{[t]}(j)$. We need a structure induction rule to formalize this fact.

Let $R(X)$ be a PDC formula, where X is a variable ranging over duration formulas. R is said to be *disjunction closed*, if $R(X \vee Y)$ is provable from $R(X)$ and $R(Y)$ assuming that $X \wedge Y \Leftrightarrow ff$.

AR 11 *Let $R(X)$ be disjunction closed.*

1. *If $R(\lceil \rceil)$ and $R(ff)$ hold, and $R(X; \lceil v \rceil^1)$ is provable from $R(X)$ for any $v \in V$, then $R(D)$ holds for any duration formula D.*

2. *If $R(\lceil \rceil)$ and $R(ff)$ hold, and $R(\lceil v \rceil^1; X)$ is provable from $R(X)$ for any $v \in V$, then $R(D)$ holds for any duration formula D.*

Using this rule, we can prove the following Markov property.

Theorem 3 *For arbitrary duration formulas D, D_1 and D_2 and any state $v \in V$, if*

$$\mu(D_1; \lceil v \rceil^1) = \mu(D_2; \lceil v \rceil^1)$$

then

$$\mu(D_1; \lceil v \rceil^1; D \wedge (\int 1 = r)) = \mu(D_2; \lceil v \rceil^1; D \wedge (\int 1 = r))$$

This says that the satisfaction probability of a formula after a state $v \in V$ only depends on the state v, but not on what has happened before v.

Theorem 4 *For any duration formulas D_1 and D_2 and nodes $v_i, v_j, v_k \in V$,*

$$\tau(v_i, v_j) * \tau(v_j, v_k) * \mu(D) = \tau(v_i, v_k) * \mu(D')$$

where

$$D \triangleq (D_1 \wedge (\int 1 = r); \lceil v_i \rceil^1; \lceil v_k \rceil^1; D_2; \int 1 = 1)$$
$$D' \triangleq (D_1 \wedge (\int 1 = r); \lceil v_i \rceil^1; \lceil v_j \rceil^1; \lceil v_k \rceil^1; D_2)$$

This theorem gives a way to calculate the probability from the middle of a behaviour.

6   Examples

6.1   A Gas Burner

Consider the Gas Burner illustrated in Section 3.1. We show how to estimate the probability of the requirement. We assume one time unit to be one second and take the result in [2],

$$(\textbf{Ds-1} \wedge \textbf{Ds-2}) \Rightarrow \textbf{Req-1}$$

Thus,

$$\neg\textbf{Req-1} \Rightarrow (\neg\textbf{Ds-1} \vee \neg\textbf{Ds-2})$$

From **AR** 3 and **AR** 4, we then have

$$\mu(\neg\textbf{Req-1}) \leq \mu(\neg\textbf{Ds-1} \vee \neg\textbf{Ds-2}) \leq \mu(\neg\textbf{Ds-1}) + \mu(\neg\textbf{Ds-2})$$

$$\mu(\neg\textbf{Ds-1}) = \mu(tt; (\lceil leak \rceil \wedge (\int 1 > 1)); tt)$$

$$\mu(\neg\textbf{Ds-2}) = \mu(tt; ((\lceil leak \rceil; \lceil \neg leak \rceil; \lceil leak \rceil) \wedge (\int 1 < 32)); tt)$$

In what follows, we present a recursive calculation of $\mu(\neg\textbf{Ds-1})$. From the duration calculus,

$$\neg\textbf{Ds-1} \wedge \int 1 \leq 1 \Leftrightarrow ff$$

Therefore, by Theorems 2.1, 1.3 and 1.1,

$$T \leq 1 \Rightarrow \mu(\neg\textbf{Ds-1}) = 0$$

Also, $(\neg\textbf{Ds-1} \wedge \int 1 = 2) \Leftrightarrow \lceil leak \rceil^1; \lceil leak \rceil^1$; but $\tau_0(leak) = 0$ thus $T = 2 \Rightarrow \mu(\neg\textbf{Ds-1}) = 0$, by Theorems 2.1, 1.3 and **AR 10** and **AR 9**.

Ds-1 is violated for the first $t + 1$ time units, $t > 1$, if and only if **Ds-1** has been violated for the first t time units already, or **Ds-1** holds for the first t time units but it is violated one time unit later. These two cases are mutually exclusive. The probability of the first case can be recursively calculated. The second case is formulated by duration formula

$$(\textbf{Ds-1}; \int 1 = 1) \wedge \neg\textbf{Ds-1}) \wedge (\int 1 = t + 1)$$

that is equivalent to

$$(\textbf{Ds-1}; \lceil \neg leak \rceil^1; \lceil leak \rceil^1; \lceil leak \rceil^1) \wedge (\int 1 = t + 1)$$

which describes how the last unit's *leak* violates **Ds-1**.

All the arguments above can be formalized in the duration calculus, but are not completely presented here. By Theorem 1 and **AR 3** and **AR 10**, the probability of the second case can be calculated recursively by the probability of $(\textbf{Ds-1}; \lceil \neg leak \rceil^1)$

$$T = t + 1 \Rightarrow$$

$$\mu(\textbf{Ds-1}; \lceil \neg leak \rceil^1; \lceil leak \rceil^1; \lceil leak \rceil^1) =$$

$$p_{12} * p_{22} * \mu(\textbf{Ds-1}; \lceil \neg leak \rceil^1; \int 1 = 2)$$

Now we introduce two functions $\mathcal{P}(t)$ and $\mathcal{Q}(t)$ defined respectively by

$$(T = t \wedge t > 2) \Rightarrow \mu(\neg \textbf{Ds-1}) = \mathcal{P}(t)$$

$$(T = t \wedge t > 2) \Rightarrow \mu(\textbf{Ds-1}; \lceil \neg leak \rceil) = \mathcal{Q}(t)$$

As discussed above, with the calculus we can prove

$$\begin{cases} \mathcal{P}(t+1) &= \mathcal{P}(t) + p_{12} * p_{22} * \mathcal{Q}(t-1) \\ \mathcal{Q}(t+1) &= p_{11} * \mathcal{Q}(t) + p_{12} * p_{21} * \mathcal{Q}(t-1) \end{cases}$$

where $\mathcal{P}(t)$ gives us the concrete value for $\mu(\neg \textbf{Ds-1})$ for a given value of t.

The calculation of $\mu(\neg \textbf{Ds-2})$ is given recursively as follows. From **AR 2**,

$$\mu(\neg \textbf{Ds-2}) = 1 - \mu(\textbf{Ds-2})$$

And in the duration calculus,

$$(\textbf{Ds-2} \wedge \int 1 > 0) \Leftrightarrow ((\textbf{Ds-2} \wedge (tt; \lceil leak \rceil^1) \vee \textbf{Ds-2} \wedge (tt; \lceil \neg leak \rceil^1)))$$

So by **AR 3** and Theorem 1.1, we have

$$(T > 0) \Rightarrow (\mu(\textbf{Ds-2}) = \mu(\textbf{Ds-2} \wedge (tt; \lceil leak \rceil^1)) + \mu(\textbf{Ds-2} \wedge (tt; \lceil \neg leak \rceil^1))))$$

Let $\mathcal{U}(t)$ and $\mathcal{V}(t)$ be the functions such that

$$T = t \Rightarrow \mu(\textbf{Ds-2} \wedge (tt; \lceil leak \rceil^1)) = \mathcal{U}(t)$$

$$T = t \Rightarrow \mu(\textbf{Ds-2} \wedge (tt; \lceil \neg leak \rceil^1)) = \mathcal{V}(t)$$

We can derive the following recursive equations for $\mathcal{U}(t)$ and $\mathcal{V}(t)$ in the calculus.

$$
\begin{cases}
\mathcal{U}(t+1) = p_{22} * \mathcal{U}(t) + \begin{cases} p_{11}^{28} * p_{12} * \mathcal{V}(t-29) & \text{if } t > 29 \\ p_{11}^{t-1} * p_{12} & \text{if } 1 \leq t \leq 29 \\ 0 & \text{if } t < 1 \end{cases} \\[4mm]
\mathcal{V}(t+1) = p_{21} * \mathcal{U}(t) + p_{11} * \mathcal{V}(t)
\end{cases}
$$

Using the above mutually recursive equations, we can calculate $\mu(\mathbf{Ds\text{-}2})$ and thus $\mu(\neg\mathbf{Ds\text{-}2})$.

As an illustration we have calculated that, to ensure that the requirement **Req-1** with a probability no less than 0.999 within one day (24 hours); the implementation components must be so chosen that the transition probabilities p_1, p_2, p_{11}, p_{12}, p_{21} and p_{22} guarantee

$$
\mu(\neg\mathbf{Ds\text{-}1}) + \mu(\neg\mathbf{Ds\text{-}2}) \leq 0.001
$$

within one day.

6.2 A Protocol Over an Unreliable Communication Medium

Consider a *medium* through which a *sender* process sends messages to a *receiver* process. To describe the behaviour of the protocol, we introduce the following states.

- s, b, m and r represent that the sender, buffer, medium and receiver are active respectively. e represents an error state of the medium.

- The protocol starts from state s, i.e. the sender is active to send a message.

- The sent message is written into a buffer within one time unit.

- The medium receives the message from the buffer within one time unit.

- Within one time unit, the message sent by the medium may be received by the receiver with probability p and the protocol enters state r, or the message is lost with probability $1-p$ and the protocol enters error state e.

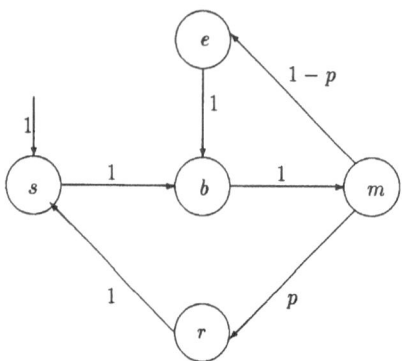

Figure 3: A Protocol Over an Unreliable Medium

- If the message is lost, within one time unit, the medium re-reads the message from the buffer.

- If the message is received by the receiver, within one time unit, the receiver acknowledges the sender and the sender is ready to send another message.

The protocol is illustrated in Figure 3. The primitive states which make up the composite states are not elaborated and the unreachable states and transitions with probability 0 are eliminated.

The first kind of properties we are interested in are the so called *soft-deadline* properties [7]. It describes that starting from the state s, within t time units, i.e. within the interval $[0, t]$, the receiver receives at least one message with probability q. This is formalized in terms of PDC as

$$T = t \Rightarrow \mu(\neg(\smallint r = 0)) = q$$

or equivalently,

$$T = t \Rightarrow \mu(\smallint r > 0) = q$$

It is not difficult to derive

$$(3k < T \leq 3(k+1)) \Rightarrow (\mu(\neg(\smallint r = 0)) = 1 - (1 - p)^k)$$

When $p = 0.9$, i.e. ten percent of the messages are lost, we have $T = 7 \Rightarrow \mu((\smallint r > 0)) = 0.99$. This gives the same result as presented in [7].

Another kind of properties is to describe the *upper bound* of error occurrences for a given interval $[0, t]$. This property can be specified by the satisfaction probability of $\int e \leq n$, and also reasoned about in the calculus.

Now let us discuss the probability of the *reoccurrence of the error state*. We define the following formulas for shorthand. Let D_1 be the duration formula

$$(\int 1 = k); \lceil e \rceil^1; \lceil \neg e \rceil \wedge (\int 1 = k_1); \lceil e \rceil; tt$$

and D_2 be the duration formula

$$(\int 1 = k); \lceil e \rceil; tt$$

The conditional probability of D_1 under D_2 defines the probability of error reoccurrence in k_1 time units. When $\mu(D_2) \neq 0$, it is equal to $\dfrac{\mu(D_1 \wedge D_2)}{\mu(D_2)}$, denoted $a(k_1)$.

Using natural induction on k_1, we can derive

$$a(k_1) = \begin{cases} 1 - p & \text{if } k_1 = 2 \\ p^{n+1} * (1 - p) & \text{if } k_1 = 4n + 6 \\ 0 & \textbf{otherwise} \end{cases}$$

by proving

$$(T \geq k + k_1 + 2) \Rightarrow (\mu(D_1 \wedge D_2) = a(k_1) * \mu(D_2))$$

7   Conclusion and Discussion

We have presented a probabilistic duration calculus based on discrete Markov processes. The main motivation for the development has been to unify the calculations undertaken in order to verify the correctness as well as reliability of a design. We have illustrated this by analysing the 0.999 probability of failure for a safety requirement **Req-1**

$$T \leq 1day \Rightarrow \mu(\textbf{Req-1}) \geq 0.999$$

for a simple Gas Burner. The analysis uses that a correct design results from two design decisions **Ds-1** and **Ds-2**. For these formulas the probabilities can be reduced to failure probabilities for components.

The starting point for the development has been the two model approach in [1]; which, however, uses continuous Markov processes. We have chosen discrete Markov processes in order to have a simpler set of rules; but it is planned to investigate a similar extension for continuous time.

The approach in [7], based on CTL in [9], can be used to analyse soft-deadline properties. It can also be used to analyse **Ds-1**; but we have not succeeded in using it to analyse probabilities of **Req-1**, **Ds-2**, error occurrences or reoccurrences.

Future work will include investigation of the incorporation of reliability analysis with design for fault-tolerant systems [13, 14, 15]. A further essential piece of work is to experiment with schemas for calculating concrete probabilities; because this would allow us to benefit from the existing knowledge about Markov processes in practical applications.

Acknowledgements

We would like to acknowledge helpful discussions with N.H. Hansen, M.R. Hansen and H. Rischel, and advice from the referees on the presentation of this paper.

References

[1] E. Sørensen, J. Nordahl, and N. Hansen, "From CSP models to Markov models: a case study," Tech. Rep. ProCos, Institute of Computer Science, Technical University of Denmark, DK-2800 Lyngby, Denmark, 1991. Accepted by *IEEE Transactions on Software Engineering*.

[2] C. Zhou, C. Hoare, and A. Ravn, "A calculus of durations," *Information Processing Letters*, vol. 40, no. 5, pp. 269–276, 1992.

[3] K. Hansen, A. Ravn, and H. Rischel, "Specifying and verifying requirements of real-time systems," *ACM SIGSOFT '91 Conference on Software for Critical Systems*, December 1991.

[4] A. Ravn and H. Rischel, "Requirements capture for embedded real-time systems," *IMACS-IFAC Symposium MCTS, Lille, France*, pp. vol.2, pp. 147–152, 1991.

[5] J. Skakkebæk, A. Ravn, H. Rischel, and C. Zhou, "Specification of embedded, real-time systems," *Proc. of 4th EuroMicro Workshop on Real-Time Systems*, (Athens, Greece), pp. 116–121, IEEE Computer Society Press, Los Alamos, California, 3-5, June 1992.

[6] P. M. Melliar-Smith and R. L. Schwartz, "Formal specification and mechanical verification of SIFT: A fault tolerant flight control system," *IEEE Trans. on Computers*, vol. 31, no. 7, 1982.

[7] H. Hansson and B. Jonsson, "A framework for reasoning about time and reliability," *Proc. of 10th IEEE Real-Time System Symposium, S:a Monica, Ca.*, 1989.

[8] K. Larsen and A. Skou, "Bisimulation through probabilistic testing, " in *Proc. of 16th ACM Symposium on Principles of Programming Languages*, 1989.

[9] E. Clarke, E. Emerson, and A. Sistla, "Automatic verification of finite-state concurrent systems using temporal logic specification: A practical approach," in *Proc. of 10th ACM Symp. on Principles of Programming Languages*, pp. 117–126, 1983.

[10] M. R. Hansen and C. Zhou, "Semantics and completeness of duration calculus," *Lecture Notes in Computer Science 600* (W.-P. d. R. J.W. de Bakker and G. Rozenberg, eds.), pp. 209–226, Springer-Verlag, 1991.

[11] W. Feller, *An Introduction to Probability Theory and Its Applications*, vol. 2. John Wiley & Sons, Inc. New York, London, 2nd ed., 1966.

[12] G. E. Hughes and M. J. Cresswell, *A Companion to Modal Logic*. London: Methuen and Co., 1984.

[13] Z. Liu, *Fault-Tolerant Programming by Transformations*. PhD thesis, Department of Computer Science, University of Warwick, Coventry CV4 7AL, UK, 1991.

[14] Z. Liu and M. Joseph, "Transformation of programs for fault-tolerance," *Formal Aspects of Computing*, vol. 4, no. 5, pp. 442–469, 1992.

[15] J. Nordahl, *Specification and Design of Dependable Communicating Systems*. PhD thesis, Department of Computer Science, Technical University of Denmark, DK-2800 Lyngby, Denmark, 1992.

Timed Statecharts and Real Time Logic

L. BARROCA

DCSC/Dept. of Computer Science

University of York

York YO1 5DD, UK

Abstract

This paper deals with simple ways of proving and deducing timing constraints of real time systems. Timed Statecharts are used to specify real time reactive systems. From such specifications assertions can be proved about time constraints and new restrictions deduced. To express such constraints Real Time Logic formulas are written directly from the Timed Statecharts. A set of rules is presented to translate Timed Statecharts constraints into RTL formulas. Two examples illustrate the work presented.

Key Words: Timing constraints, Timed Statecharts, RTL, verification of safety constraints.

1 Introduction

The specification of real-time systems has been the subject of thorough research recently and several notations have been proposed to express both the real-time constraints of a system and to prove that safety constraints are observed. The main interest of the work developed here has been to apply some existing languages and to express behaviour together with time constraints and be able, within the same framework, to prove properties about the specification.

Several languages for real-time systems are based on transition systems. A system is characterised with a behaviour that is defined by a set of possible *states*, and it evolves between states by the occurrence of either external or internal *events*. Transitions between states occur under certain conditions and they provoke the execution of actions. The work done with Statecharts [1, 2, 3, 4] has been closely looked at, since it not only provides the elements to describe the behaviour of *reactive* systems[1], but also has the visual presentation that enables specifications to be more intuitively analysed.

2   Timed Statecharts

Statecharts have been proposed by Harel [1] as an extension of state machines and state diagrams for the specification of *reactive* systems. The main extensions that this visual formalism proposes to the traditional state/transition diagrams are clustering (depth), orthogonality (concurrency), and they also support refinement and broadcast communication (communication between concurrent state machines).

Statecharts have been recently extended, in a simple way, to deal with time constraints – Timed Statecharts [5]. *Lower* and *upper* time bounds can be attached to each transition; they represent the minimum and maximum time that the transition can be enabled before it is triggered. Timed Statecharts are a graphical representation of Timed Transition Systems[2] [6, 7]. Timed Transition Systems have been extended to deal with both discrete and continuous time; these are called Hybrid Systems and are extensively discussed in [8]. In this paper the discussion will be limited to systems where time is discrete.

In Timed Statecharts transitions happen instantaneously; they are represented by arcs between states. A label is attached to a transition having, in general terms, the *name* of the transition (if any), the *event* that triggers the transition, a *condition* (a predicate), and an *action* to be taken when the transition is activated.

name: event[condition]/action

When a transition is triggered by the occurrence of an event it happens immediately, when the event occurs, without any time bounds. For time bound transitions the label has the following form:

---

[1]*Reactive* systems are characterised by reacting to stimuli from the environment; their behaviour is conditioned by the occurrence of events.

[2]A Timed Transition System is defined as $S = < V, \Sigma, \Theta, \Gamma, l, u >$, where V is the set of variables, Σ the set of states, Θ the set of initial states, Γ the set of transitions, l minimal delay, and u maximal delay.

name: [condition][lower_bound, upper_bound] /action

Conditions and actions are expressions in terms of the variables of the system. Actions create events or assign values to shared variables. Any part of the label can be omitted.

Timed Statecharts are asynchronous in the sense that there is no mechanism to force concurrent components to evolve together or not at all. However, an event generated by a transition triggers all the transitions that can be taken at that moment in time and have that event as the triggering event. All these immediate transitions occur before time progresses.

The semantics of Timed Statecharts have been described in [6].

A few small examples will be shown using the Timed Statecharts formalism, and the timing constraints of these systems will be translated into RTL. The derivation of RTL formulas will be shown in a set of small simple rules.

2.1   Representing Periodic or Non-Periodic Events

The model proposed for Timed Statecharts represents directly the occurrence of non-periodic events. However, if we want to represent events that occur periodically, this can be done implicitly. As we will see in the next sections we can express timing constraints on transitions and relate consecutive occurrences of an event.

If a clock has to be represented explicitly it will be represented by a process that will periodically change, as represented in Figure 1. An event can be generated when each transition occurs (*down* or *up* and this event can trigger periodic transitions).

3   Real Time Logic

Real Time Logic is a formal language that captures time constraints of real time systems. It was introduced by [9] to deal with absolute time as opposed to relative temporal order. RTL has no modal operators. Its basic concepts are *actions, state predicates, events and timing constraints*. Actions are marked by events. Events are either Start Events (↑*action*, marking the beginning of an action), Stop Events (↓*action*,

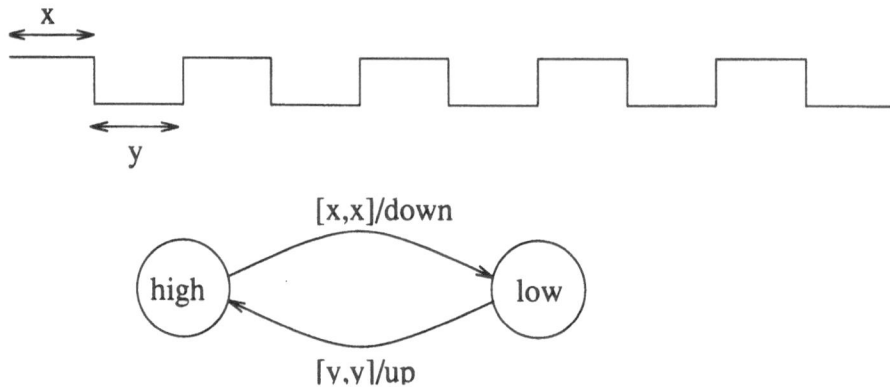

Figure 1: Timed Statechart for a periodic clock.

marking the end of the action), Transition Events (*S:=T*, *S:=F*, marking a change of state), or External Events (ΩE). An important notion, in RTL, is the *occurrence function* which captures the notion of time, $@(E,i)$—the ith occurrence of event *E*. Predicates are constructed stating values about attributes of the system. Formulas in RTL use universal and existential quantifier and first order logic. As pointed out in [10], RTL equations are usually expressed as inequalities imposing a lower and upper bound on occurrences of events. The similarity with lower and upper bounds in Timed Statecharts has motivated this study.

The appeal of RTL is its pragmatic nature. Since our objective is to explore ways of combining precision with ways of presenting formalism to non-formal people, RTL is an attractive notation to be used to prove properties about systems. The examples described here will explore ways of combining the visual presentation of Timed Statecharts with RTL to state and prove assertions about the behaviour of a real time system.

4   Translating Timed Statecharts into Real Time Logic

Timed Statecharts will be used to show the behaviour of real time systems, stating assertions about time constraints. They are intuitive to use and rather expressive; ideally, the process involving the proof of time constraints and the enforcement of relations between time bounds should also be done in an intuitive and pragmatic way (where real time in seconds can be expressed).

A set of rules will be presented to translate the Timed Statecharts time bounds into RTL formulas. As was mentioned before, both Timed Statecharts and RTL express time through occurrences of events. In Statecharts events cause transitions; in RTL events determine the start and end points of actions, and the change of predicate values.

The proof methodology proposed in [11] for transition systems will be followed closely for the definition of the rules. However, we allow explicit reference to time and express the real time properties using absolute time instead of eventuality and relative temporal order.

The labels in the Timed Statechart transitions represent a step that can be timed or not. Only the timed steps are relevant for the timing constraints. A transition relates two states and a label. The timing constraints that will need to be proved include lower and upper time bounds (real-time response, min/max separation of related endpoints). A transition τ will be seen as a transformation between two states (characterised by predicates p and q) — $\{p\}\ \tau\ \{q\}$.

RTL formulas establish occurrences of events. The correspondence that will be established will be based on the following assumptions:

1. States will be represented by the start and end events that correspond to entering and exiting the state; we will use the notation $@(p := T, i)$ for the event corresponding to state p being entered, and $@(p := F, i)$ for the event associated with leaving state p.

2. The RTL inequalities used are of the form
   $@(Ev_1, i) + c_1 \leq @(Ev_2, j) \leq @(Ev_1, i) + c_2$;
   this type of formula has the following implicit quantification
   $\forall i, t\ \exists j, t' \bullet @(Ev_1, i) = t \land @(Ev_2, j) = t' \land t + c_1 \leq t' \leq t + c_2$;

3. In Timed Statecharts, a (non-event triggered) transition becomes enabled when the predicate asserting that the initial state for that transition has been entered and the condition in the label becomes true;
   $@(enabled(\tau) := T, i) = @((p \land c) := T, i)$
   In a timed transition, we consider that a transition has been enabled for a certain amount of time t, if since the state has been entered time t has elapsed and c, the condition on the label, has been continuously

true during that amount of time. This is expressed imposing that if the transition is taken, only then $(p \wedge c)$ become false.
$$@(enabled(\tau) := F, i) = @((p \wedge c) := F, i)$$

4. An event triggered transition is enabled when the triggering event occurs, after the starting state for that transition has been entered. An event triggered transition is immediately triggered when it becomes enabled.
$$@(p := T, i) \leq @(enabled(\tau) := T, i) \wedge$$
$$@(enabled(\tau) := T, i) = @(Ev, j) \wedge$$
$$@(Ev, j) = @(p := F, i)$$

5. The occurrence of a transition is marked by the postcondition becoming true. In RTL terms this can be expressed by
$$@(enabled(\tau) := F, i) = @(q := T, j).$$

6. The *bounded response* asserts the maximum time within which something will happen
$$@(enabled(\tau) := T, i) + u \geq @(q := T, j)$$

7. The *bounded invariance* asserts the minimum time during which there will not be a change in some assertion.
$$@(enabled(\tau) := T, i) + l \leq @(q := T, j)$$

The underlying model of time [6] assumes a totally ordered domain that can be either represented by the natural or the real numbers. The behaviour of a system is represented by a timed trace and between two consecutive situations either the state is the same or time is the same. The assumption that time and state do not change simultaneously is an abstraction that avoids ambiguities about the exact time at which a state change has occurred.

4.1   Translation Rules

To extend the proofs for every real-time transition, different possible situations have to be considered: the single transition, the sequencing of transitions, two transitions leading to a same state, and transitions occurring in parallel. The following rules apply to time bound transitions; when the time interval, on a transition label is omitted it corresponds to $[0, 0]$.

The rules establish always the conditions under which the transitions occur. The quantification will be assumed to be as shown above, when nothing else is said.

1. *lower and upper bound single-step* rule (Figure 2) — this is the direct application of the *bounded response* and *bounded invariance*.
   $\{p\}\ \tau\ \{q\}$

   (1) $@(enabled(\tau) := T, i) + l_1 \leq @(q := T, j) \leq @(enabled(\tau) := T, i) + u_1$

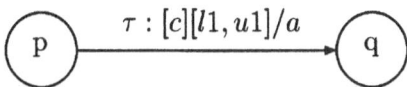

   Figure 2: Single-step Timed Transaction.

2. *lower and upper bound transitivity* rule (Figure 3)
   $\{p\}\ \tau_1\ \{q\}\ \tau_2\ \{r\}$

   When a transition follows another this rule establishes the time bound of their sequential composition. If condition c_2 is a predicate other than *true* then the inequality in (3) only the leftmost inequality can be enforced — $@(enabled(\tau_1) := T, i) + l_1 + l_2 \leq @(r := T, l)$.

   (1) $@(enabled(\tau_1) := T, i) + l_1 \leq @(q := T, j) \leq @(enabled(\tau_1) := T, i) + u_1$
   (2) $@(enabled(\tau_2) := T, k) + l_2 \leq @(r := T, l) \leq @(enabled(\tau_2) := T, k) + u_2$
   (3) $@(enabled(\tau_1) := T, i) + l_1 + l_2 \leq @(r := T, l) \leq$
       $@(enabled(\tau_1) := T, i) + u_1 + u_2$

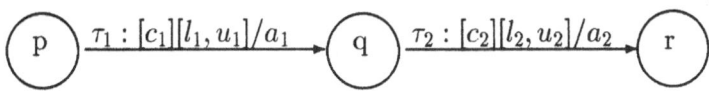

   Figure 3: Transitivity of Timed Transactions.

3. *lower and upper bound disjunctive* rule (Figure 4)
   $\{p_1\}\ \tau_1\ \{q\} \vee \{p_2\}\ \tau_2\ \{q\}$
   In case two transitions are simultaneously enabled one of them will

be taken non-deterministically with the following time constraint.

(1) $@(enabled(\tau_1) := T, i) = @(enabled(\tau_2) := T, k) = t$
(2) $t + l_1 \leq @(q := T, j) \leq t + u_1 \vee t + l_2 \leq @(q := T, j) \leq t + u_2$
(3) $t + min(l_1, l_2) \leq @(q := T, j) \leq t + max(u_1, u_2)$

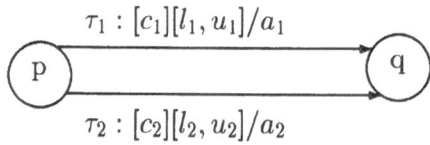

Figure 4: Disjunctive Timed transactions.

4. *lower and upper bound parallel* rule (Figure 5)
$\{p_1\} \ \tau_1 \ \{q_1\}$
$\{p_2\} \ \tau_2 \ \{q_2\}$
Up to now we have dealt with a single process; here, we consider how the timing constraints can be considered when we have two concurrent processes. Synchronisation in Timed Statecharts is simulated by the generation of events that trigger transitions in concurrent active machines. Such transitions take place before time progresses.

(1) $@(enabled(\tau_1) := T, i) + l_1 \leq @(q_1 := T, j) \leq @(enabled(\tau_1) := T, i) + u_1$
    $\wedge @(enabled(\tau_2) := T, k) \leq @(q_1 := T, j) \leq @(enabled(\tau_2) := F, k)$
(2) $@(enabled(\tau_2) := F, k) = @(q_2 := T, l) = @(q_1 := T, j)$

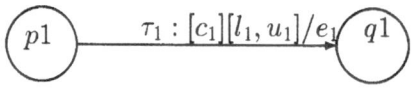

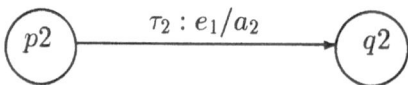

Figure 5: Parallel Timed Transactions.

5 Examples

5.1 Gate in a Rail Crossing

The first example that we will show is a simple rail road crossing which has been presented in [12, 13, 14]. The system described is a monitor for a gate in a rail crossing. There are two entities: the monitor that receives information about the situation of a train on the line, and a gate controller that controls the position and movement of the gate.

The behaviour of each of these entities will be described using a Timed Statechart. The monitor can be in the following states: *Ap-*

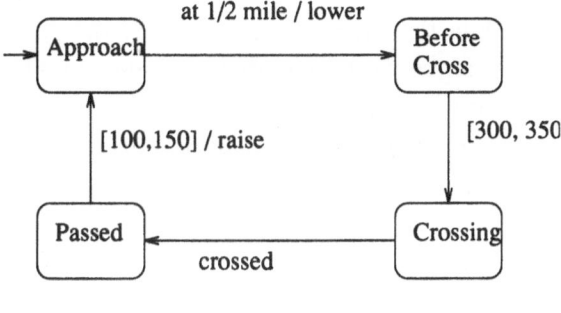

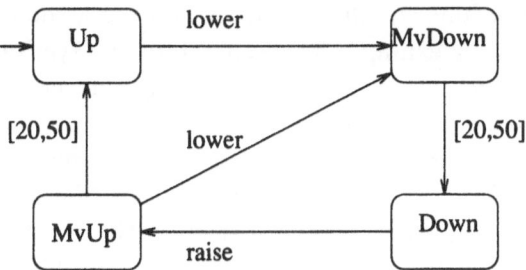

Figure 6: Timed Statecharts for the Monitor and Gate Controller

proach (there isn't a train in less than half a mile distance), *Before Cross* (there is a train within half a mile), *Crossing* (there is a train crossing), *Passed* (the train has crossed). The transitions between states occur by events related to the monitoring of the train distance (Train at half a mile, Train crossed) and with time delays. When *Train at half a mile* event happens the transition between *Approach* and *BeforeCross* occurs immediately; the transition between *BeforeCross* and *Crossing* occurs after a delay between 300 and 350 units of time; the transition from *Crossing* to *Passed* is triggered by the *Train crossed* event, and, after a delay between 100 and 150 units of time, the transition to *Approach* will occur. When transition from *Approach* to *BeforeCross* occurs the event *lower* is generated, and when the transition from *Passed* to *Approach* occurs the event *raise* is generated.

The gate controller can be in the following states: *Up, MvDown, Down, MvUp*. When the event *lower* occurs (generated by the monitor) the transition from *Up* to *MvDown* is triggered; the gate has to be down within a timeout of 50 (having taken at least 20 units of time); it will start moving up when the event *raise* occurs. If the gate is moving up and it receives another request to go down (event *lower*) a transition to *MvDown* occurs immediately.

Initially both states *Approach* and *Up* are entered.

5.2   Two Machines and a Buffer

This example has been presented in [6]. Two machines communicate by a buffer that can only hold one item at a time. The buffer gets an item from one machine, to the other. A label is associated with each transition; *set-up*, for example, is the name of the transition from *Idle* to *Busy* in *M1* which has to occur between the lower and upper time bounds of *l3* and *u3*. When transition from *m2* to *p2* occurs (between the lower and upper time bounds of *l4* and *u4*) in the *Buffer*, the event *ready* is created; this triggers the transition from *Idle* to *Busy* in *M2* (if *M2* is in state *Idle*) which creates consequently the event *take* which will trigger the transition from *p2* to *m1* in the Buffer. This is an example of how broadcast communication occurs between concurrent machines. Initially the two machines are in state *Idle* and the Buffer is in state *p1*.

A situation that should never occur in this system is the machine *M1* outputting an item (issuing a *put* signal) when the buffer is in the *moving* state (see Figure 7) We will try to prove which relation should be established between the time bounds so that this situation never happens. The figure shows the Timed Statecharts for this problem.

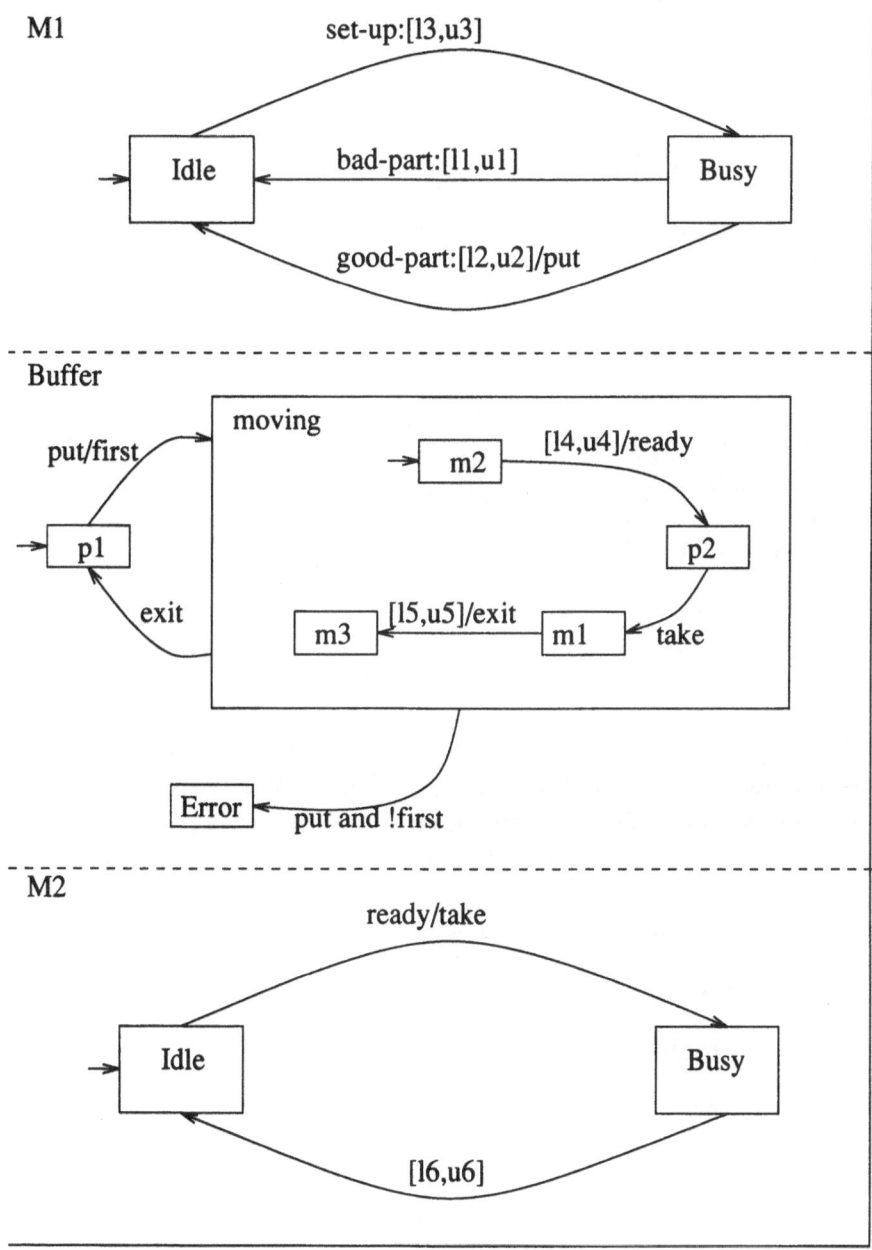

Figure 7: Timed Statecharts for the machines and buffer problem.

6   Proving Safety Constraints

In this section the rules stated above will be applied to prove properties
of the systems and impose time constraints.

6.1   Rail Road Crossing

In this problem there is a safety assertion that has to be established between the time constraints that guarantees that there is no train crossing
and the gate is up. We will show that the time bounds on the transitions
verify the safety constraint.

$\forall i \; \exists j \bullet @(Crossing := T, i) \geq @(Down := T, j) \land$
$\qquad @(Crossing := F, i) \leq @(Down := F, j)$

(1) $@(BeforeCross := T, i) = @(MvDown := T, j)$ Rule 4
(2) $@(MvDown := T, j) + 20 \leq @(Down := T, j) \leq @(MvDown := T, j) + 50$
(3) $@(MvDown := T, j) + 300 \leq @(Crossing := T, i) \leq @(MvDown := T, j) + 350$
(4) $@(Crossing := T, i) > @(Down := T, j)$

(1) $@(Passed := F, i) = @(Down := F, j)$ Rule 4
(2) $@(Crossing := F, i) = @(crossed, k)$
(3) $@(crossed, k) + 100 \leq @(Passed := F, i) \leq @(crossed, k) + 150$
(4) $@(Crossing := F, i) < @(Down := F, j)$

We have applied the rules for parallel composition of two processes,
the bounded response, and bounded invariance. We have proved that the
time constraints specified for the monitor and the gate controller verify
the safety assertion. For any further refinement it has to be proved
that these constraints are still true. If, for example, we allow for some
delay on communications (on message passing) the relation between
these constraints will still have to hold.

6.2   Two Machines and a Buffer

The timing constraints, in the Timed Statechart for the problem of the
two machines and the buffer (see Figure 7), will also be represented by
a set of RTL formulas. We will first analyse the parallel execution of
machine *M1* and the *Buffer*, and then of the *Buffer* and machine *M2*.

The transition (labelled *good-part*) from *Busy* to *Idle* in the machine
M1 generates the event *put*. We want to establish such a relation between the time bounds that the interval between two consecutive occur-

rences of the event *put*, in the machine *M1*, has to allow an *exit* event to occur in the Buffer in between. We relate any occurrence (*i*) of the event *put* with its next (*i+1*) occurrence. By the *transitivity* rule, the event *exit* occurs in a time bounded by the occurrence of the transitions from *m2* to *p2*, *m1*, *m3*, in the *Buffer*.

Applying the same rule twice we get the inequalities between *l3+l2* and *l4+l5*, and between *u3+u2* and *u4+u5*. The transition from *p2* to *m1* in the buffer occurs instantaneously, as soon *p2* is entered; once the transition from *m2* to *p2* creates the event *ready* and the transition triggered by this event in *M2* creates the event *take*. These two transitions occur before time progresses.

(1) $@(put, i) = @(m2 := T, i)$
(2) $@(put, i) + l4 \leq @(p2 := T, i) \leq @(put, i) + u4$ Rule 1
(3) $@(p2 := T, i) = @(take, i) = @(m1 := T, i)$ (transitions are inst.)
(4) $@(m1 := T, i) + l5 \leq @(exit, i) \leq @(m1 := T, i) + u5$ Rule 1
(5) $@(put, i) + l4 + l5 \leq @(exit, i) \leq @(put, i) + u4 + u5$ Rule 2
(6) $@(put, i) + l3 + l2 \leq @(put, i + 1) \leq @(put, i) + u3 + u2$ Rule 2
(7) $@(put, i + 1) > @(exit, i)$ Requirement
(8) $@(put, i) + l3 + l2 > @(put, i) + u4 + u5$
(9) $l3 + l2 > u4 + u5$

If this relation holds between the time bounds of the different transitions, then the *Error* state should never be entered. However, we have to guarantee that machine *M2* is idle when the event *ready* is generated by the *Buffer*, otherwise the expression (3) above will not be true. This is a synchronisation requirement between the *Buffer* and machine *M2*. We do this by imposing restrictions on the time bounds of machine *M2*. Two consecutive transitions to *p2* in the *Buffer* have to allow for the upper bound *u6* in the time that machine is *Busy*.

(1) $u6 \leq @(p2 := T, i + 1) - @(p2 := T, i)$ Requirement
(2) $@(put, i + 1) + l4 \leq @(p2 := T, i + 1) \leq @(put, i + 1) + u4$ Rule 1
(3) $@(put, i) + l3 + l2 + l4 \leq @(p2 := T, i + 1) \leq$
         $@(put, i) + u3 + u2 + u4$ (6) above
(4) $@(put, i) + l4 \leq @(p2 := T, i) \leq @(put, i) + u4$ Rule 1
(5) $l3 + l2 \leq @(p2 := T, i + 1) - @(p2 := T, i) \leq u3 + u2$ ((3),(4))
(6) $u6 \leq l3 + l2$

If these relation between the time bounds holds we guarantee that the machine *M2* will be available when it is needed.

6.3   Considerations on Proofs

We have established the translation from Timed Statecharts to RTL in terms of the occurrences of events, and occurrences of state change. The existence of procedures to analyse a certain subset of RTL formulas is discussed in [10]. This subset is formed by the arithmetic inequalities involving two occurrence functions and an integer constant. The RTL formulas that we have used in this paper are included in this subset.

The rules are based on the semantics of Timed Statecharts (transitions being instantaneous and the persistence of events on a step until time progresses) and on the underlying model of time previously referred to (time change and state change do not occur simultaneously, and time progresses). It is clear that if this work is to be of some real use there will definitely be the need for tool support in the application of the rules.

6.4   Related Work

The work that we present in this paper has been mainly inspired by the work presented in [11]. We wanted to experiment with the extension of the proof rules proposed for Timed Transition Systems and define a new set of rules that could be applied to Timed Statecharts and expressed in RTL. The choice of RTL was done mainly by its pragmatic nature and the possibility of expressing absolute time which is required to express some safety constraints properties.

There is however, a close similarity with both Modecharts [15] and with the Timed Transition Model [16, 17]. Modecharts are based on the concept of *mode*; a mode is taken from Parnas notation [18], and it stands for a partition on the state space. This work also uses RTL to express the timing constraints. The Timed Transition Model uses RTTL (a temporal logic). Both these models present decision procedures for the verification of certain class of properties and they are both based on the construction of computation or reachability graphs.

The comparison with both Modecharts and Ostroff's Timed Transition Model can be made in terms of the way the timing properties are reasoned about. Both the Modecharts and the TTM approach make use of a reachability graph that has to be constructed before any properties can be proved. We avoid the need of the graph (which is quite impractical to build for systems with several transitions) by defining a set of rules that can be used in the combination of several transitions and by reasoning about the inequalities imposed in the RTL formulas; we use the laws of algebraic inequality. Also, we think that the approach of replacing the notion of state by the notion of mode (as done in Modecharts) does not always lead to the most intuitive transition diagrams; it forces the consideration of not only change of mode but also change of state variables. The advantage of using Real Time Logic as opposed to other temporal logics is the possibility of dealing explicitly with absolute time and in a pragmatic way. RTL also allows for the representation of intervals.

In the Timed Statechart approach we have the possibility of expressing synchronisation by broadcast communication which does not seem feasible in Modecharts. The rules defined allow the definition of constraints on several Statecharts in parallel.

We have not explicitly addressed how these rules will cope with the decomposition of states in Statecharts. However, if inter-level transitions are not allowed (as proposed in [19]) the application of the rules to superstates or decomposed states can be done similarly.

7   Conclusions

A natural way to define reactive systems is through transition systems; for real time reactive systems the equivalent are Timed Transition Systems. The formalism of Timed Statecharts has been used to specify the behaviour and the timing of a system. They are a visual representation of Timed Transition Systems. The visual formalism makes the specification much clearer and intuitive. Properties about such systems can be proved from their specification when the Timed Statecharts are translated into Real Time Logic formulas. They combine the pragmatic way of dealing with concrete time with rigour of proofs. It was shown that a simple set of rules can be applied to each type of transition to prove timing relations. The examples given have shown the usability of such a procedure.

It is hoped that this technique can be used in increasing the reliability of systems without losing sight of the fact that the specification and the formal verification should be accessible and reasonable to use in large systems and understandable to systems engineers who do not necessarily have the full technical background. The automation of the process of generating the RTL formulas from the Timed Statecharts should be taken now as the next step so that tools being developed for RTL can be used here.

Acknowledgements

This work is part of the Dependable Computing Systems Centre, at the University of York, funded by British Aerospace.

References

[1] D. Harel, "Statecharts: A visual formalism for complex systems," *Science of Computer Programming*, vol. 8, pp. 231–274, 1987.

[2] D. Harel, "On visual formalisms," *Communications of the ACM*, vol. 16, no. 4, pp. 514–530, 1988.

[3] D. Harel, H. Lachover, A. Naamad, and A. Pnueli, "Statemate: A working environment for the development of complex reactive systems," *IEEE Transactions on Software Engineering*, vol. 16, no. 4, pp. 403–414, 1990.

[4] D. Harel, A. Pnueli, J. Schmidt, and R. Sherman, "On the formal semantics of statecharts," *2nd IEEE Symposium on Logic in Computer Science*, (New York), IEEE, 1987.

[5] T. Henzinger, Z. Manna, and A. Pnueli, "Timed transition systems," *Proc. of the REX Workshop–Real-Time: Theory and Practice*, 1991.

[6] Y. Kesten and A. Pnueli, "Timed and hybrid statecharts and their textual representation," *Formal Techniques in Real Time and Fault Tolerant Systems, LNCS 571* (J. Vytopil, ed.), pp. 591–620, Springer Verlag, 1991.

[7] O. Maler, Z. Manna, and A. Pnueli, "From timed to hybrid systems," 1992. presented at the School on Formal Techniques in Real-Time and Fault-Tolerant Systems, Nijmegen, The Netherlands.

[8] O. Maler, Z. Manna, and A. Pnueli, "A formal approach to hybrid systems," *Proc. of the REX Workshop–Real-Time: Theory and Practice*, 1992.

[9] F. Jahanian and A. K. Mok, "Safety analysis of timing properties in real-time systems," *IEEE Transactions on Software Engineering*, vol. SE-12, no. 9, pp. 890–903, 1986.

[10] F. Jahanian and A. K. Mok, "A graph-theoretic approach for timing analysis and its implementation," *IEEE Transactions on Computers*, vol. C-36, no. 8, pp. 961–975, 1987.

[11] T. Henzinger, Z. Manna, and A. Pnueli, "Temporal proof methodologies for real-time systems," *Proc. of 18th ACM Symp. Princ. of Prog. Languages*, pp. 353–366, ACM, 1991.

[12] F. Jahanian and D. Stuart, "A method for verifying properties of Modechart specifications," *Proc. of 9th Real Time Systems*, pp. 12–21, 1988.

[13] N. Leveson and J.Stolzy, "Safety analysis using Petri Nets," *IEEE Transactions on Software Engineering*, vol. SE-13, no. 3, pp. 386–397, 1987.

[14] J. Ostroff, *Temporal Logic for Real-Time Systems*. Research Studies Press Ltd, 1989.

[15] F. Jahanian, R. Lee, and A. K. Mok, "Semantics of Modechart in Real Time Logic," *Proc. of 21st Annual Hawai International Conference on System Science*, pp. 479–489, 1988.

[16] J. Ostroff, "Automated verification of timed transition models," *Automatic Verification Methods for Finite State Machines, LNCS 407* (J. Sifakis, ed.), Springer-Verlag, 1989.

[17] J. Ostroff, "Verification of finite state real-time embedded computer systems," *9th IEEE Int. Conf. on Distributed Computing Systems*, pp. 207–216, IEEE, 1989.

[18] D. Parnas, K. Heninger, J. Kallender, and J. Shore, "Software requirements for the A-7E aircraft," memorandum report 3876, Naval
Research Laboratories, Washington, D.C., 1978.

[19] F. Maraninchi, "Argonaute: Graphical description, semantics and
verification of reactive systems by using a process algebra," *Automatic Verification Methods for Finite State Machines, LNCS 407*
(J. Sifakis, ed.), Springer-Verlag, 1989.

Fault-Tolerant Distributed Sort Generated from a Verification Proof Outline*

H. LUTFIYYA

Department of Computer Science

University of Western Ontario

London, ONTARIO N6A 5B7 CANADA

M. SCHOLLMEYER, B. McMILLIN

Department of Computer Science

University of Missouri-Rolla

Rolla, MO 65401 USA

Abstract

The problem of *executable assertion generation* is at the core of providing operational fault tolerance for *distributed memory systems*. The application-oriented fault tolerance paradigm provides a basis for selection of assertions through examining the "Natural Constraints" of the problem. A set of high-level properties are used to extract assertion generating properties from the problem specification. The next logical step is to consider formal methods to help in automated generation of executable assertions. While formal methods have been applied to the sequential/shared memory programming environment, the distributed memory environment presents special challenges. In this paper, we describe *Changeling*, which provides a systematic approach, based on the mathematical model of program verification, to deriving executable assertions that can be evaluated in the faulty distributed computing environment. We apply the approach to parallel bitonic sort as a model problem.

Key Words: Executable Assertions, Formal Methods, Bitonic Sort, Concurrent Program Verification, Fault Tolerance, Changeling.

---

*This work was supported in part by the National Science Foundation under Grant Number MSS-9216479, and, in part, from the Air Force Office of Scientific Research under contract number F49620-92-J-0546.

1 Introduction

It is important for both life-critical, and non-life-critical distributed systems to meet their specification at run time [1]. Large, complex, distributed systems, are subject to individual component failures which can cause system failure. Fault tolerance is an important technique to improve system reliability. The fault detection aspect identifies individual faulty components (processors) before they can affect, negatively, overall system reliability.

A failure occurs when the user observes that a resource does not perform as expected. The failure is the result of some part of the resource entering a state which is contrary to the specification of the part. The cause of the resource entering such a state is referred to as a fault. When a system can recover from a fault without exhibiting a failure, then the system has fault tolerance. Reliability is a measure of the probability that a specific resource will perform a required function for a specified period of time, usually the item's life time, even in the presence of faults. The higher the probability the higher the reliability of the system is considered to be.

Many methodologies for improving system reliability have been developed throughout the years. These different methodologies fall into two basic groups: fault masking techniques and concurrent techniques. Early attempts at improving system reliability used fault-masking methods; these methods make the hardware tolerant of faults through the multiplicity of processing resources. In contrast, concurrent fault detection methods attempt to locate component errors which can lead to system failure. Once the faults are identified, reconfiguration and recovery [2] are used to deal with the fault. This paper focuses on detecting the occurrence of errors. Recovery and reconfiguration are different issues. Work in concurrent detection methods includes self-checking software [3] and recovery blocks [4], which instrument the software with assertions on the program's state, watchdog processor [5], which monitors intermediate data of a computation, and algorithm-based fault tolerance [6] which imposes an additional structure on the data to detect errors. These methods define structure for fault tolerance, but do not, generally, give a methodology for instantiating the structure.

Application-oriented fault tolerance [7], by contrast, provides a heuristic approach, based on the "Natural Constraints," to choosing executable assertions from the software specification. These executable assertions [3], in the form of source language statements, are inserted into a program for monitoring the run-time execution behavior of the program. The general form is as follows:

$$\textbf{if} \neg ASSERTION \textbf{ then } ERROR$$

Executable assertions are used to ensure that the program state, in the actual run-time environment, is consistent with the logical state specified in the assertion; if not, then an error has occurred and a reliable communication of this diagnostic information is provided to the system such that reconfiguration and recovery can take place. The heuristics for selection of the actual executable assertions are based on three metrics of *progress*, *feasibility*, and *consistency*.

What our earlier work lacks is a theoretical foundation built upon mathematical models and theories. In general, theoretical foundations can provide (1) criteria for evaluation, (2) means of comparison, (3) theoretical limits and capabilities, (4) means of prediction, and (5) underlying rules, principles, and structure. This paper describes *Changeling* as a formal method using the mathematical model of axiomatic program verification to construct executable assertions for error checking in distributed systems. Application of the *Changeling* system is a two step process. First, from a verification proof outline, *Changeling* converts a shared memory proof outline into a distributed memory proof outline, which closely matches the distributed operational environment. Second, *Changeling* transforms the assertions from the proof outline into executable assertions.

This paper is organized as follows. In Section 2 we provide a brief overview of concurrent axiomatic proof systems which form the basis for our GAA system. Section 3 describes application-oriented fault tolerance and *Changeling*. Section 4 presents some details of the verification proof of a parallel bitonic sort algorithm. Section 5 describes how the parallel bitonic sort algorithm is made error detecting and Section 6 analyzes the error coverage of the error-detecting algorithm.

2 A Short Background on Program Verification

The axiomatic approach to program verification is based on making assertions about the program variables before, during and after program execution. These assertions characterize specific properties of interest about program variables and relationships between them at various stages of program execution. Program verification requires proofs of theorems of the following type:

$$< P > S < Q > \tag{1}$$

where P and Q are assertions, and S is a statement or sequence of statements of the language. The interpretation of the theorem is as follows: if P is true before the execution of S and if the execution of S terminates, then Q is true after the execution of S. P is said to be the *precondition* and Q the *postcondition* [8]. P and Q are also referred to as program *specifications*.

In program verification logical assertions are determined that describe the effect of each statement that comprises program S. Each of these logical assertions must hold if the cumulative effect of these statements results in the execution of S satisfying the logical assertion Q. This collection of assertions is normally referred to as the "verification proof outline".

As a model axiomatic proof system, consider the work of [9], which is used for Hoare's model of concurrent programming, Communicating Sequential Processes (CSP) [10]. In this system, the first step is to prove appropriate properties about the individual processes in isolation. The individual statement axioms are omitted here except for the communication and parallel inference axioms.

The parallel inference rule is the following:

$$\frac{(\forall i :< P_i > S_i < Q_i >) \wedge \ satisfied \ and \ interference - free}{< (\forall i : P_i) > [||_{i=1:n} \, \rho_i : S_i] < (\forall i : Q_i) >} \tag{2}$$

The parallel rule (where A_i is the process label of the process that contains S_i) implies that construction of the proof of a parallel program can be derived from the partial correctness properties of the sequential programs that comprise it. However, the sequential components of the parallel program contain communication commands. The communication axiom is as follows:

$$< P > \alpha < Q > \tag{3}$$

where α is a communication command.

The communication axiom implies that full concurrent proofs of the appropriate properties of individual processes require assumptions to be made about the effect of the communication commands. A "satisfaction proof" is then used to show that these assumptions are "legitimate". Let us examine the proof outline of the matching communication pair [1].

$$\rho_1 :: [... < P_1 > \rho_2?x < Q_1 > ...] \tag{4}$$

$$\rho_2 :: [... < P_2 > \rho_1!y < Q_2 > ...] \tag{5}$$

The effect of these two communication commands is to assign y to x. This implies that $Q_1 \wedge Q_2$ is true after communication if and only if

$$(P_1 \wedge P_2) \rightarrow (Q_1 \wedge Q_2)_y^{x} \text{ [2]}. \tag{6}$$

A "satisfaction proof" is such that the above is proven for every matching communication pair. This is called the *rule of satisfaction.*

The proof system in [9] (GAA proof system) makes use of global auxiliary variables (GAVs). Auxiliary variables may affect neither the flow of control nor the value of any non-auxiliary variables. Otherwise, this unrestricted use of auxiliary variables would destroy the soundness of the proof system. Hence, auxiliary variables are not necessary to the computation, but they are necessary for verification. Auxiliary variables are used to record part of the history of the communication. Shared reference to auxiliary variables allows for assertions relating the different communication histories. This requires a proof of "non-interference". Any update of a global auxiliary variable on any processor is assumed to be immediately known to all processors.

There are other axiomatic proof systems. The proof system in [11] uses local auxiliary variables. The proof system in [12] uses communication sequences. A communication sequence for process ρ_i is the sequence

---

[1]The notation $\rho_i?x$ in CSP notation denotes receiving a message into variable x from process ρ_i. Correspondingly, $\rho_j!y$ denotes sending a message with value y to process ρ_j.

[2]This stands for the predicate $Q_1 \wedge Q_2$ with all instances of y replaced by x

of all communications that process ρ_i has so far participated in. Each
process ρ_i has a variable denoting its communication sequence, which
is updated for each communication. This allows for proof rules that
can make inferences about the communications sequences. Thus, it is
sufficient to do only sequential proofs of each component process in a
parallel program. Our work in [13] shows that these three proof systems
are equivalent in that they allow for the same properties to be proven.
However, it is easier to reason in some systems more than others.

3   Changeling and Application-Oriented Fault Tolerance

Application-oriented fault tolerance works on the principle of testing
at run time the intermediate logical assertions from the verification proof
outline i.e. application-oriented fault tolerance works on the following
principle:

> *If we test and ensure intermediate results of a program's com-
> putation meet its specification, the end solution meets its spec-
> ification if the intermediate results meet their specification. If
> processor errors occur that do not affect the solution, then they
> are not errors of interest. Program verification provides these
> tests.*

The above principle yields a formal statement of application-oriented
fault tolerance; we generate the executable assertions from the logical
assertions used in the verification proof outline of $< P > S < Q >$. The
executable assertion generated corresponding to any logical assertion Q_i
from the verification proof outline is the following:

if $\neg Q_i$ then $ERROR$

Formally, this ensures that if P is true before the concurrent program
S begins execution, S tests at run time that S satisfies the specification
as defined by P and Q, by using the embedded executable assertions
generated from the assertions of the verification proof. Conversely, the
assertions of the verification proof represent the properties that must be
satisfied by the run-time environment; an error that causes the execution

of the program not to satisfy the specified assertions will be flagged as an error by the executable assertions.

The reader may be suspicious that some program S may be changed into a program S' by an error that satisfies the specification as defined by P and Q. Consider, as an example, a program S computing some value x with postassertion $< Q > \equiv < x \geq 0 >$. Suppose that S should compute $x = 3$. A program S' may actually compute $x = 4$. The postcondition is still satisfied, although, the value is not what was intended. This is not a problem with the validity of the postassertion, it is a weakness of the specification. If $x = 3$ was what was really intended, then the proper postassertion should have been $< Q > \equiv < x = 3 >$. If $< Q > \equiv < x \geq 0 >$ is a sufficient specification for the application at hand, then there is no problem.

To eliminate confusion between the testing of intermediate results (via logical assertions) for correctness with respect to the algorithm and the evaluation of the executable assertions derived from the verification proof in the run-time environment, we will refer to the former as the *verification environment* and the latter as the *(distributed) operational environment*.

To summarize, the transformation of an algorithm to an error-detecting algorithm involves using the assertions of the verification proof as executable assertions that are to be embedded into the algorithm.

Taking an application from the verification environment to the distributed operational environment is not a straightforward task. It is this difficulty that inspired the development of *Changeling*. *Changeling* consists of four distinct components:

1. The GAA Proof System described in Section II

2. An HAA proof system which mimics closely the distributed operational environment

3. Formal conversion from GAA to HAA

4. Formal translation of assertions in the HAA proof system to executable assertions and reducing state information to improve run-time efficiency

These components are described in the following paragraphs.

3.1   History of Auxiliary Variable (HAA) Verification System

The logical assertions from the GAA verification environment cannot be directly used as executable assertions in the distributed environment; in the distributed environment, there are no global variables. Thus, to evaluate, at run time, logical assertions containing global auxiliary variables, an explicit updating mechanism must be created. Here we develop the verification proof system (HAA) in which updates of global auxiliary variables are exchanged at communication time. This matches, more closely, the operational environment. We show that every verification proof outline in the GAA proof system has the same properties in the HAA proof system, i.e., satisfaction and non-interference; thus, implying that the HAA proof system has the soundness and completeness properties of the original GAA proof system. The existence of the HAA proof system allows for proofs that can be directly transformed to executable assertions in the run-time environment.

Developing the HAA system requires us to keep track of which processes communicate with which other processes. Each process needs to record its global auxiliary variable updates with respect to all other processes. When communication occurs between two processes, they need to exchange the updates and locally apply them (the updates). This is formalized in the following definitions.

Definition 3.1 *For a process ρ_i, h_i denotes the sequence of all communications that process ρ_i has so far participated in as the receiving process. Thus, h_i is a list consisting of tuples (these are different from the [12] tuples; all future reference to tuples will refer to the following tuples) representing matching communication pairs of the form*

$$[\rho, (Var, Val), T, C]$$

where ρ is a process from which ρ_i receives from, Var is the variable that ρ is transmitting to ρ_i with formal parameter Val. T denotes the time at which the value Val was assigned to variable Var and C denotes the communication path.

In Section 3.2, we will use the path notion.

Since we have several processes running in parallel and there exists no concept of a global time, the time T is a local time represented by an instantiation counter that is incremented by one after every execution of a statement. This permits an ordering (time-stamping) for all updates of the GAVs within each process.

To be able to account for the different operations performed on the auxiliary variables, each process has to keep a history of variable updates with respect to the last communication with the other processes. These variable sets are described using the subscript of the corresponding process.

Definition 3.2 *Let g_{ij} depict the GAV set in process ρ_i with respect to process ρ_j, i.e., g_{ij} contains the changes that were made to the GAVs in ρ_i since the last communication with ρ_j. G_i is the set of sets g_{i0}, g_{i1}, ..., $g_{i(N-1)}$ in process ρ_i. Thus, when two processes ρ_i and ρ_j communicate, the values of their respective subsets, $g_{ij} \in G_i$ and $g_{ji} \in G_j$, are exchanged.*

When two processes ρ_i and ρ_j communicate, where ρ_j is the sender, ρ_j will augment the communication by sending the values of global auxiliary variables that ρ_j updated, or received updates of, between the last and current communications between ρ_i and ρ_j. We batch the changes made to the local copies of the global auxiliary variables by ρ_j since the last communication (with any other processor) in g_{jj}. Before a communication, the function ψ applies changes to all g_{jk}'s and g_{jj} is reset to null to collect future changes. Definition 3.4 formally describes the communication of g_{ji}. Definition 3.5 formally describes how process ρ_i updates G_i based on the communicated g_{ji} after communication has taken place.

Definition 3.3 *The actual set of GAVs to be sent during a communication between ρ_i and ρ_j, where ρ_j is the sender, is determined based on the variables in g_{jj}, i.e., all the variables that were updated in ρ_j since the last communication with any process. The set g_{jj} is updated every time an assignment to a GAV takes place in ρ_j and reset at communication time to the empty set. The following function $\psi(G_j, g_{jj})$ describes the update of all variable histories before a communication.*

$$\rho_j \; : \quad (\forall k, 0 \le k \le N-1)(\forall g_{jk} \in G_j)$$
$$[if \; k \ne j \; then \; g_{jk} \leftarrow g_{jk} \cup g_{jj} \; else \; if \; k = j \; then \; g_{jj} \leftarrow \emptyset]$$

The following definition formally defines the semantics of global auxiliary variable communication.

Definition 3.4 *The* primary *communication is a matching communication pair for the exchange of variables between processes which are not GAVs. It can be described by a tuple $[\rho_j, (Var, Val), t, j]$ where $t = T_j$ is the current value of the local time. It is easy to see that all communications in the GAA system are primary since GAVs are updated globally. An* augmented *communication permits the exchange of the GAVs after a primary communication occurs. In an augmented communication, the values of g_{ji} are marshalled into a message sent to process ρ_i.*

For each process ρ_i after an augmented exchange with ρ_j, ρ_i updates its set of GAVs in G_i with the new values received. This interchange is described in Figure 1 for two processes ρ_i, ρ_j and one matching communication pair within the execution sequence of the two processes.

Definition 3.5 *The updates performed in the different processes are described by a function $\phi(G_i, g_{ji})$ on the set of the GAV history and the variables to be updated. The actual update function ϕ is now defined on all the subsets within G_i on tuples of the form $[\rho_j, (g_{ji}, gvar_j), T, j]$.*

$$\rho_i \;:\; (\forall k, 0 \leq k \leq N-1)(\forall g_{ik} \in G_i)$$
$$[if\ k \neq j\ then\ g_{ik} \leftarrow g_{ik} \cup gvar_j$$
$$else\ if\ k = j\ then\ apply(gvar_j); g_{ij} \leftarrow \emptyset\,]$$

When processes ρ_i and ρ_j communicate, all old values in the set g_{ij} will be replaced by the new variables. Additionally, these new values (from $gvar_j$) are unmarshalled and applied to update the local values of process ρ_i. In this way, communication propagates GAV updates throughout the concurrent program.

It can be seen that the so-called "global auxiliary variables" in the HAA system are not really global in the sense that all processes have the same values of the variables at all times. Indeed, it is likely that at the end of the process execution some processes that ran in parallel will have different values within their set of GAVs. We show that because of non-interference, this is not a problem with respect to the proof system.

Within a process execution, two communicating processes can have arbitrary interleavings of their statements up to the communication,

For process P_i:
/* execute arbitrary set of statements excluding communication but
including assignments to auxiliary variables */
$S_{i1}; < T_i := 1 >$
$S_{i2}; < T_i := T_i + 1 >$
...;
$S_{ik}; < T_i := T_i + j >$
/* update the auxiliary variables */
$G_i \leftarrow \psi(G_i, g_{ii}); < T_i := T_i + 1 >$
/* perform communication with process P_j;
the first communication represents the actual communication */
/* the next two communications represent the exchange augment of the auxiliary variables */
$P_j ? V ; < T_i := k >$
$P_j ? g_{ji} ; < T_i := k + 1 >$
$P_j ! g_{ij} ; < T_i := k + 2 >$
/* update the auxiliary variables */
$G_i \leftarrow \phi(G_i, g_{ji}); < T_i := k + 3 >$

For process P_j:
/* execute arbitrary set of statements excluding communication
but including assignments to auxiliary variables */
$S_{j1}; < T_j := 1 >$
$S_{j2}; < T_j := T_j + 1 >$
...;
$S_{jk}; < T_j := T_j + 1 >$
/* update the auxiliary variables */
$G_j \leftarrow \psi(G_j, g_{jj}); < T_j := T_j + 1 >$
/* perform communication with process P_i;
the first communication represents the actual communication */
/* the next two communications represent the exchange augment of the auxiliary variables */
$P_i ! V ; < T_j := k >$
$P_i ! g_{ji} ; < T_j := k + 1 >$
$P_i ? g_{ij} ; < T_j := k + 2 >$
/* update the auxiliary variables */
$G_j \leftarrow \phi(G_j, g_{ij}); < T_j := k + 3 >$

Figure 1: An HAA proof outline for one matching communication pair.

but are conceptually synchronized at the communication point. The assertions will not interfere with each other due to the non-interference property of the GAA system which provides for arbitrary execution orders. Since two (or more) processes will only change (write onto) the same global auxiliary variable if they have to communicate with each other, they will also exchange other variables/data in that process and the values of the auxiliary variables will be available for the other process at the critical point: right after a communication takes place. Thus, sending only the history of the global variable updates instead of immediately providing the other process(es) with the latest information will not cause any problems, since the values of the variables will be available at the communication points, where they are in fact provided.

An example of three possible process execution sequences that are subject to non-interference is shown in Figure 2. For any two processes, non-interference will guarantee that the execution order of the two processes or any arbitrary interleaving of them will not invalidate the assertions made on the respective process statements.

Non-interference and the rule of satisfaction can be used to show that the soundness and completeness properties of the original GAA system will hold in the new HAA proof system.

Theorem 3.1 [14] *The history of auxiliary variables approach (HAA) retains the properties of the global auxiliary variables approach (GAA).*

3.2   Reliable Communication of State Information

The HAA proof system provides for direct transformation of assertions from the verification environment into executable assertions for the non-faulty distributed operational environment. However, we are concerned with the distributed faulty environment. Thus, it is necessary to ensure that faulty processors cannot fool executable assertions by incorrect augmented communication of $g's$ through sending inconsistent messages to different processors. It is necessary for this to be detected. This is the purpose of *consistency* executable assertions. Mathematically, this can be described as follows:

Definition 3.6 *For a non-faulty process ρ_i, if there exist any two tuples t_1, $t_2 \in h_i$ such that*

$$t_1 = [j, (Var, Val_1), T, C_1]$$

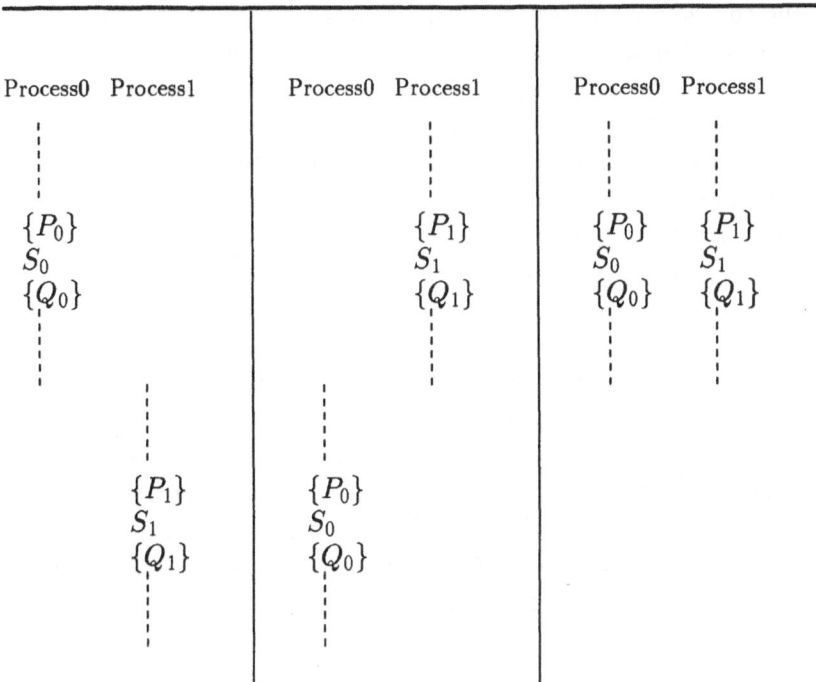

Figure 2: Some possible process execution sequences before communication takes place.

$$t_2 = [j, (Var, Val_2), T, C_2]$$

then if $Val_1 \circ Val_2$ the system is said to be inconsistent otherwise the system is said to be consistent.

$\circ$ is defined as a set of functions such that each $\circ' \in \circ$ is of functionality $dt \rightarrow T, F$ where dt is an abstract data type. Examples of $\circ'$ are $\neq$, $\subseteq$, $\neg prefix$, or some other operator appropriate to the choice of the data type of Var. Where no ambiguity results, we will refer to a particular $\circ'$ simply as $\circ$.

The strongest motivation for the consistency condition is to supplement the power of the executable assertions derived from the HAA system. When the value of a variable computed in time T is communicated to a set of processors on more than one path, there will be two or more tuples in h_i that satisfy the precondition. Under a bounded number of faults, the consistency definition of 3.6 ensures that a non-faulty processor receives a consistent set of input values for its executable assertions, otherwise, $Val_1 \circ Val_2$, and an inconsistent system can be detected. The degree of fault tolerance is based on standard network flow arguments and is not repeated here. It should be noted that all faults in communication links are mapped to a processor, thus it is enough to assume only faulty processors.

Consistency does not have to be explicit. In other words, an error-detecting program may have to explicitly add code to implement consistency. This can be done in many ways. There are classes of problems that have the property of natural redundancy in the problem variables. This implies that there are types of errors, which if they occur at stage i, eventually, at some stage j (where $j > i$), we have that stage j satisfies the properties as defined by the intermediate assertions of a verification proof, despite the fact that the error had occurred in stage i. If a program variable is naturally redundant then this means that this program variable can be constructed from other variables.

3.3   Run-Time Efficiency Considerations

The transformation from the HAA verification environment to the operational environment described above is optimal in the sense that all violations of the program's specification (in terms of the postconditions on each statement and within the limits of consistency) are caught under a bounded number of faults. However, when run-time efficiency is considered, not all of these assertions, nor all of the communicated GAVs are necessary. These two aspects of reducing complexity are treated as follows:

- Assertions involving *local variables* to a particular process which are necessary in the verification environment are useless in the distributed operational environment. Since the unit of failure and reconfiguration is at the processor level, a processor cannot be trusted to diagnose itself as faulty or fault-free. Thus, assertions using only local variables incur a run-time overhead that is not necessary and all such assertions can be deleted.

- The fault coverage of certain assertions using the GAVs may be *subsumed*. Thus, many of the remaining assertions may be removed as well. Likewise, removing some of the assertions may result in certain GAVs no longer being required. Furthermore, certain assertions may be too expensive to evaluate in the operational environment and may be deleted for that reason.

3.4　Section Summary

This section showed how the properties of the GAA system are preserved when a conversion from this formal system into another occurs; non-interference is the critical point that assures that satisfaction, soundness and completeness can be retained in the new system. The new system is important since it simplifies the operations on sets of executable assertions which are used to provide error detection in a distributed operational environment. We then showed how the HAA system can be transformed, through consistency, into an operational system and commented on run-time efficiency.

We applied this transformation to several concurrent applications ranging from concurrent database transactions schedules [15], to concurrent branch and bound [16] and obtained performance and error coverage data on each. The next section presents example applications of this system including its verification, transformation, measures of its efficiency, and error coverage.

4　Verification Details of Bitonic Sort

This section describes verification details of a bitonic sort algorithm. The target multicomputer interconnection topology considered is the popular hypercube topology. These systems can grow to over 1000 processors. In general, the topology of an n-dimensional hypercube is a graph $H(P, E)$ with $N = 2^n$ vertices called nodes labeled $P_0, P_1, \ldots, P_{N-1}$. An edge $e_{i,j} \in E$ connects P_i and P_j if the binary representations of i and j differ in exactly 1 bit. If we let this bit position be k, then $P_i = P_{j \oplus 2^k}$. Thus, in an n-dimensional hypercube, each processor connects to n neighboring processors. An actual hypercube multicomputer contains an additional host node which is not represented in H. The host has connections to the nodes mainly used for program/data downloading and result uploading. The algorithm used in this paper as a parallel sorting algorithm was introduced by Batcher in 1968 [17]. This bitonic sort algorithm was introduced as a parallel sorting algorithm that can take advantage of interconnection topologies such as the perfect shuffle and the hypercube. There exists a bitonic sort algorithm that maps directly to a hypercube topology.

The postassertion Q of any sorting is defined in the following:

Definition 4.1 *Given an input list* $\mathbf{I} = (I_i)$, *i=0,...,N-1 a sorting procedure S finds a permutation* $\Pi = (\pi)$ *such that:*

$$I_{\pi_i} \leq I_{\pi_{i+1}}, i = 0, ..., N - 2$$

or

$$I_{\pi_i} \geq I_{\pi_{i+1}}, i = 0, ..., N - 2$$

The general idea of a bitonic sort is to build up longer bitonic sequences which eventually lead to a sorted sequence.

Definition 4.2 *A* **bitonic sequence** *is a sequence of elements* O_0 ,O_1 ,... ,O_{N-1} *such that*

1. *There exists a subscript i,* $0 \leq i \leq N - 1$ *such that* $O_0 \leq O_1 \leq ... \leq O_{i-1}$ *and* $O_i \geq O_{i+1} \geq ... \geq O_{N-1}$

2. *There exists a subscript i,* $0 \leq i \leq N - 1$ *such that* $O_0 \geq O_1 \geq ... \geq O_{i-1}$ *and* $O_i \leq O_{i+1} \leq ... \leq O_{N-1}$

The fundamental operation in a bitonic sort is the compare-exchange operation, either min(x,y) or max(x,y).

Lemma 4.1 [17] *Given a bitonic sequence* $I_0 \leq I_1 \leq ... \leq I_{N/2-1}$ *and* $I_{N/2} \geq I_{N/2+1} \geq ... \geq I_{N-1}$, *each of the subsequences formed by the compare-exchange steps:*

$$min(I_0, I_{N/2}), min(I_1, I_{N/2+1}), ..., min(I_{N/2-1}, I_{N-1}).$$

and

$$max(I_0, I_{N/2}), max(I_1, I_{N/2+1}), ..., max(I_{N/2-1}, I_{N-1})$$

is bitonic with the property that $O_i \leq O_j$ *for all* $i = 0, 1, ..., N/2 - 1$ *and* $j = N/2, N/2 + 1, ..., N - 1$

Procedure Bitonic Sort;

$ia_{node} = a;$

$a_{node} = a;$

$pa_{node} = a;$

{ **for** i:=0 **to** n-1 **do**

$< Loop_i >$

   $pa_{node} = a;$

   $a'_{node} = a;$

   { **for** j:=i **downto** 0 **do**

   $< Loop_j \wedge Loop'_j \wedge a'_{node} = a_{node} >$

     { $d:=2_j;$

     $< Loop'_j \wedge a'_{node} = a_{node} \wedge d = 2^j >$

     if ($node$ **mod** (2d)¡d)

     $< Loop'_j \wedge a'_{node} = a_{node} \wedge d = 2^j >$

      { **read** into data from $node + d$;

      $< Loop'_j \wedge a'_{node} = a_{node} \wedge d = 2^j \wedge data = a_{node+d} >$

      if ($node$ **mod** $2^{i+2} < 2^{i+1}$)

      $< Loop'_j \wedge a'_{node} = a_{node} \wedge d = 2^j \wedge data = a'_{node+d} \wedge node$ **mod** $2^{i+2} < 2^{i+1} >$

       { b := max(data,a);a := min(data,a); }

       $< Loop'_j \wedge d = 2^j \wedge data = a'_{node+d} \wedge node$ **mod** $2^{i+2} < 2^{i+1}$

       $\wedge b = max(a'_{node+d}, a'_{node}) \wedge a = min(a'_{node+d}, a'_{node})$

       $\wedge \forall l, SC^S_{j,node} \leq l \leq SC^E_{j,node} \ [a \leq max(a'_l, a'_{l+d})] >$

      else

      $< Loop'_j \wedge a'_{node} = a_{node} \wedge d = 2^j \wedge data = a'_{node+d}$

      $\wedge \neg(node$ **mod** $2^{i+2} < 2^i + 1) >$

       { b := min(data,a);a := max(data,a); }

       $< Loop'_j \wedge; d = 2^j \wedge data = a'_{node+d} \wedge \neg(node$ **mod** $2^{i+2} < 2^{i+1})$

       $\wedge b = min(a'_{node+d}, a'_{node}) \wedge a = max(a'_{node+d}, a'_{node})$

       $\wedge \forall l, SC^S_{j,node} \leq l \leq SC^E_{j,node} \ [a \leq min(a'_l, a'_{l+d})] >$

      write from b to $node+d$;

     else /* Send to neighbor - we are inactive this iteration */

     /* Proof outline is symmetrical */

     { **write** from a to $node$-d; **read** into a from $node$-d; }

      $a_{node} = a;$

      $a'_{node} = a_{node};$

   } /* End for j */

} /* End for i */

Figure 3: Bitonic Sort Instrumented with Verification Assertions

For this presentation, for simplicity, we assume that $N = 2^k$ for some k and therefore the initial midpoint is $N/2$. Since each compare-exchange involves only a comparison between elements whose subscripts differ on only one bit and the number of elements is always 2^k, if we have one element per processor, then the bitonic sort can be easily implemented on a hypercube of dimension $n = log_2 N$ [18]. As a notational convenience, we define the following:

Definition 4.3 *The home subcube $SC_{i,j}$ of dimension i of a processor P_j is the subcube of size 2^i that begins with processor P_k, $k = j - (j \bmod 2^i)$ and includes all processors through P_l, $l = j - (j \bmod 2^i) + 2^i - 1$. Let $SC_{i,j}^S$ denote the index k and $SC_{i,j}^E$ denote the index l.*

The Bitonic sort algorithm, instrumented with the assertions necessary for verification, is shown in Figure 3 for each node *node*. Of particular interest are assertions $Loop_i$ which asserts that at each execution of the outer iteration (a *stage*), a bitonic sequence in the subcube of dimension $SC_{i,node}$ is made.

$Loop_i$ states that the values of the local variables a after each loop execution produces a bitonic subsequence in each subcube of size 2^{i+1}. $Loop_i$ also states that the values of the local variables a after each loop execution are a permutation of the previous values of the local variables a. Formally, $Loop_i$ is the following:

$$i \neq 0 \rightarrow (node \bmod 2^{i+1} < 2^i \rightarrow a_{SC_{i,node}^S} \leq \ldots \leq a_{SC_{i,node}^E} \wedge$$
$$node \bmod 2^{i+1} \geq 2^i \rightarrow a_{SC_{i,node}^S} \geq \ldots \geq a_{SC_{i,node}^E}) \wedge$$
$$\exists l, 0 \leq l \leq N - 1 a_{node} = ia_l \wedge \exists l, SC_{i,node}^S \leq l \leq SC_{i,node}^E a_{node} = pa_l$$

and $Loop_j$ is displayed in Figure 4.

The assertion $Loop_j'$ is constructed by replacing in the assertion $Loop_j$ each a_i with a_i'.

a_{node}, a_{node}' and ia_{node} where $0 \leq node \leq N - 1$ are the auxiliary variables used in the proof outline. The postcondition is that the output list is a permutation of the input list and that it is sorted. To decrease complexity the precondition is assumed to include that $n > 1$. In order to make use of this simplification, it is assumed that no faulty behavior occurs in the execution of the first iteration of the outer loop. This helps in eliminating tedious but trivial cases.

$\forall m, j \leq m \leq i - 1$

$\quad \big[node \bmod 2^{i+2} < 2^{i+1} \rightarrow$

$\quad\quad \forall k, SC^S_{m+1,node} \leq k \leq SC^E_{m+1,node} \ \forall l, SC^S_{m+1,node+2^{m+1}} \leq l \leq SC^E_{m+1,node+2^{m+1}} [a_k \leq a_l] \ \wedge$

$\quad\quad node \bmod 2^{i+2} \geq 2^{i+1} \rightarrow$

$\quad\quad \forall k, SC^S_{m+1,node} \leq k \leq SC^E_{m+1,node} \ \forall l, SC^S_{m+1,node-2^{m+1}} \leq l \leq SC^E_{m+1,node-2^{m+1}} [a_k \geq a_l] \big) \ \wedge$

$\quad i = j \ \rightarrow$

$\quad\quad \big(node \bmod 2^{i+2} < 2^{i+1} \ \rightarrow \ a^S_{SC_{i,node}} \leq \cdots \leq a^E_{SC_{i,node}} \ \vee$

$\quad\quad node \bmod 2^{i+2} \geq 2^{i+1} \ \rightarrow \ a^S_{SC_{i,node}} \geq \cdots \geq a^E_{SC_{i,node}} \big) \ \wedge$

$\quad node \bmod 2^{i+2} < 2^{i+1} \rightarrow$

$\quad\quad \big(\forall l, SC^S_{i,node} \leq l < SC^E_{i,node} [a_l \leq a_{l+1}] \ \vee$

$\quad\quad \exists l, SC^S_{i,node} \leq l < SC^E_{i,node} \forall k, SC^S_{i,node} \leq k < l [a_k \geq a_{k+1}] \ \vee$

$\quad\quad \forall k, l \leq k < SC^E_{i,node} [a_k \geq a_{k+1}] \big) \ \wedge$

$\quad node \bmod 2^{i+2} \geq 2^{i+1} \rightarrow$

$\quad\quad \big(\forall l, SC^S_{i,node} \leq l < SC^E_{i,node} [a_l \geq a_{l+1}] \ \vee$

$\quad\quad \exists l, SC^S_{i,node} \leq l < SC^E_{i,node} \forall k, SC^S_{i,node} \leq k < l [a_k \leq a_{k+1}] \ \vee$

$\quad\quad \forall k, l \leq k < SC^E_{i,node} [a_k \leq a_{k+1}] \big) \ \wedge$

$\exists l_{0 \leq l \leq N} a_{node} = i a_l \ \vee \ \exists l_{SC^S_{i,node} \leq l \leq SC^E_{i,node}} a_{node} = p a_l$

Figure 4: The inner loop invariant $Loop_j$

It is assumed for the sake of simplicity that we have loop synchronization. This simplifies proofs of non-interference. As part of inner loop synchronization, assign a_{node} to a'_{node}. This is done before any check for loop invariants. Loop invariants are checked at beginning of loops, but loop indices are changed at end. This implies that the loop indices are changed at the end of a loop iteration before invariants are checked.

5 Deriving the Executable Assertions

The executable assertions are either executed after the receipt of a message or after an execution of the outer loop of the program is completed.

The verification proof showed the naturally expected result; at termination, bitonic sort returns a sorted permutation of the original input. This required showing at each level of recursion (1) bitonic sequences are maintained, (2) each sorted sequence is a permutation of the previous bitonic sequence. In the loop formulation, two loop invariants resulted from the verification; the outer loop showed properties (1) and (2) directly while the inner loop assertion supports the loop's contribution to the outer loop.

In conversion from the verification proof in GAA to HAA, g_{ij} for any two processors k and l is

$$g_{kl} = \{\{ia_m\}, \{pa_m\} \mid 0 \le m \le N - 1\}$$

We now turn our attention to observations on efficiency of the transformed program. We can apply heuristics to reduce the overhead penalty of the error-detecting algorithm.

Even after deleting all assertions that use only local variables, we are left with redundant assertions. Consider the inner and outer loop invariants; each of these use global auxiliary variables. If the inner assertion fails, then the outer assertion also fails. Thus, for efficiency, we can delete the inner assertion from the transformed program.

The auxiliary variable pa_l does not need to be communicated directly within the augmented communication. At the end of each iteration of the outer loop, we may simply assign $pa_l \leftarrow ia_l$ since the current bitonic sequence values become the old bitonic sequence values in the next iteration. Thus, we can modify g, as follows

$$g'_{kl} = \{\{ia_m\}, \mid 0 \le m \le N - 1\}$$

For consistency, since we have a regular, point-to-point interconnection graph, the natural communication patterns define for each processor the frequency of received input values (global auxiliary variables) for each executable assertion. In the bitonic sort algorithm, a processor will receive two copies of a global auxiliary variable before an executable assertion. The rest of this section shows the following (1) a processor l will collect in stage $i + 1$ the local copies of a computed at the end of stage i from processors in the subcube $SC_{i+1,l}$ (Theorem 2) and (2) processor l receives two values of a_k for each k in $S_{i+1,l}$ (Theorem 3).

None of the processes have the "global picture" of the values of the auxiliary variables. Therefore, it becomes necessary to communicate these values. This is done by augmenting values of the auxiliary variables at the end of stage i, in communication that occurs naturally at stage $i + 1$. Mathematically, this can be described as follows:

Definition 5.1 *Define* λ_{ij}^l, *where* $i > 0$, *as follows:*

If $l \bmod 2^{j+1} < 2^j$ *then*

$$\lambda_{ij}^l = \{l, l + 2^j\} \qquad j = i$$
$$\lambda_{ij}^l = \lambda_{i,j+1}^l \cup \lambda_{i,j+1}^{l+2^j} \qquad 0 \le j < i$$

and **if** $l \bmod 2^{j+1} \ge 2^j$ **then**

$$\lambda_{ij}^l = \{l, l - 2^j\} \qquad j = i$$
$$\lambda_{ij}^l = \lambda_{i,j+1}^l \cup \lambda_{i,j+1}^{l-2^j} \qquad 0 \le j < i$$

Definition 5.1 states that λ_{ij}^l reflects the natural flow of augmented communications that processor l receives at time (i, j), where i and j are the values of the outer and inner loop indices, respectively.

It is shown in Theorem 2 that the values of the local variables a computed in stage $i - 1$ and collected by processor l in stage i are from the processors in the subcube denoted by $SC_{i+1,l}$. In other words, λ_{i0}^l denotes the processors that processor l collected values of the local variables, a, computed in stage $i - 1$ of the subcube denoted by $SC_{i+1,l}$.

Theorem 2 is proven by showing that (1) For processor l, the number of processors from which processor l collects from in stage i is the same number of processors in the subcube denoted by $SC_{i+1,l}$ and (2) that the processors from which processor l collects from in stage i is a subset of the processors in the subcube denoted by $SC_{i+1,l}$.

The number of processors from which processor l collects from in stage i is the same number of processors in the subcube denoted by $SC_{i+1,l}$. It is trivial to show that the processors from which processor l collects from is the same as the processors in the subcube denoted by $SC_{i+1,l}$. [19]

Theorem 5.1 $\lambda_{i0}^l = SC_{i+1,l}$.

Proof: $\lambda_{ij}^l \subseteq SC_{i+1,l}$ and $|\lambda_{ij}^l| = 2^{i-j+1}$; therefore $|\lambda_{i0}^l| = 2^{i+1}$ from above. Definition 4.3 implies that $|SC_{i+1,l}| = 2^{i+1}$. It can now be concluded that $\lambda_{i0}^l = SC_{i+1,l}$. $\square$

¿From Theorem 2 and Definition 5.1 it can be concluded that processor l receives the local copies of the variable a of the subcube $SC_{i+1,l}$ computed by the end of stage $i - 1$.

It is entirely possible for an arbitrary faulty processor to send different versions of the same message to different processors. Hence, it is
necessary for the error-detecting algorithm to ensure that every processor receives the same version of a message. This can be accomplished
by sending multiple copies of each element via vertex disjoint paths.
It must be remembered that if $k \bmod 2 < 1$ then processor k knows
processor $k + 1$'s local copy of a at the end the end of stage i (a dual
statement can be said for the case $k \bmod \geq 1$. Therefore, at the beginning of stage $i + 1$, processor k will communicate its local copy of a and
the local copy of a corresponding to processor $k + 1$. Observing that if j
is not 0 then for any processor $l \in SC_{i+1,k}$, k and $k + 1$ cannot be both
be in λ_{ij}^l.

Theorem 5.2 *For any $k \in \lambda_{i0}^l$, where $k \bmod 2^{j+1} \leq 2^j$ and $i > 0$ there
exists n, p, where $n \neq p$ and $a_k \in g_{nl}$, $a_k \in g_{pl}$.*

Proof. It is easy to show from Definition 4.3 that if $i > 0$ then k and
$k + 1$ are members of λ_{i0}^l. Therefore, since k and $k + 1$ are members of
λ_{i0}^l, but for $j > 0$, k and $k + 1$ are not simultaneously in λ_{ij}^l, then we
have that there exists n, p, $n \neq p$ and $a_k \in g_{nl}$, $a_k \in g_{pl}$. $\square$

6   Error Coverage and Complexity

This section discusses the error coverage and complexity of the bitonic
sort algorithm embedded with the executable assertions that is used to
create an error-detecting algorithm.

Lemma 6.1 *The executable assertions at stage $i+1$ detect a non-bitonic
sequence in a subcube of size 2^{i+2} and detects whether the local copies of
the variable a in each subcube of size 2^{i+2} of stage i is a permutation of
local copies of the variable a in subcubes of size 2^{i+1} of stage $i - 2$.*

Proof. From Theorem 2, processor l receives the local copies of the variable a computed by the end of stage i from processors in $\lambda_{i+1,0}^l$ in stage
$i + 1$. From Theorem 3, we have that the processors $\lambda_{i+1,0}^l$ is equivalent to the processors in the subcube denoted by $SC_{i+2,l}$. Definition 4.3
implies that $|SC_{i+2,l}| = 2^{i+2}$. The verification proof showed that the lo-
cal copies of the variable a after each loop execution produces a bitonic
subsequence in each subcube of size 2^{i+1}. Therefore, $SC_{i+2,l}$ contains a

bitonic sequence of length 2^{i+2}. Since, these values are collected in stage $i+1$, the executable assertions can only detect a non-bitonic sequence at this point. Similar statements can be said about the permutation of values. $\Box$.

Lemma 6.2 *At the end of stage i, at least one processor in $SC_{i-1,j}$ can detect an error made by processor P_j that results in either (1) a non-bitonic subsequence in the subcube containing j of size 2^i or (2) if the maximum number of faulty nodes in $SC_{i-1,j}$ is 1, the ability to determine if the local copies of the variable a computed in stage $i-2$ is a subset of the local copies of the variables a computed in stage $i-1$. It is assumed that the local copies of the variable a that are computed in stage $i-2$ are error free.*

Proof. The proof is by induction on i. For i=0, each pair of processors $l, l+1$, where l is even, contains the actual correct initial values a_l, a_{l+1}. This forms a complete bitonic sequence of length 2.

Assume that the local copies of the variable a computed in stage $k-1$ have been verified to be correct with respect to the local copies of the variable a computed in stage $k-2$. By Definition 5.1, the local copies of the variable a computed in stage k and collected by processor l in stage $k+1$ is denoted by $\lambda^l_{k+1,0}$ and the local copies of the variable a computed in stage $k-1$ and collected by processor l in stage k is denoted by λ^l_{k0}. Since each local copy of a collected by processor l in stage $k+1$ is reported through two vertex disjoint paths, by Theorem 3, the effects of a single faulty relay are limited to one of these paths. If the two candidate elements differ, an error is signaled. Thus, if the sender is faulty, it must send identical values along both paths. If the consistency condition is met, then by Lemma 2, an error will be flagged if there is a non-bitonic sequence in the subcube containing j of size 2^k or if the values of the local copies of the variable a computed in stage $k-2$ are not a subset of the values of local copies of the variable a computed in stage $k-2$. $\Box$

Theorem 6.1 *The error-detecting algorithm produces either a correct bitonic sort or stops with an error in the presence of at most one faulty node.*

Proof: By application of Lemma 3, at each step i of the outer loop, each bitonic sequence is verified. The final extra stage verifies that last sequence. Since we are allowed 1 faulty node per $SC_{i,j}$, a processor P_j can detect any faulty behavior. $\square$

The parallel bitonic sort algorithm introduced in Section 4, has a time and communication complexity of $O(log_2^2 N)$.

The local copies of a collected from the other processors are stored in one dimensional arrays, where the ith element is the value of the local variable a in stages $i - 1$ and $i - 2$, respectively, resulting in an error detecting algorithm of time complexity of $O(N)$ and communication complexity of $O(log_2^2 N + Nlog_2 N)$ [7]. As expected, there is a performance penalty to pay for the increased reliability. However, it has been shown that this extra cost does result in a more efficient algorithm than the alternative of reliable sequential sorting.

7   Summary

This paper has shown how to translate a verification proof for a parallel bitonic sort algorithm into an error-detecting algorithm. It is easy to see that executable assertions are directly derivable from the assertions of the verification proof, while the addition of consistency executable assertion is derived from the need to ensure that each processor has a consistent view of the auxiliary variables used in the assertions of the verification proof.

The translation required the choice of some subset of assertions from the verification proof. One major factor in the choice of the assertion was the granularity. Since processors may only communicate via messages, assertions requiring a great deal of extra communication in order to communicate the values of the variables of that assertion are undesirable; it is more desirable to choose assertions in which the communication of values of the variables of that assertion dooes not require extra communication complexity. The assertions chosen will subsume the assertions not chosen for executable assertions. In other words, if the intermediate results violate assertions not chosen for executable assertions, then the intermediate results will violate those assertions that were chosen for use as executable assertions. On the other hand, it is desirable to check for faults as often as possible in order to decrease the amount of time it takes to detect faults. Further research will examine this problem.

References

[1] J. Laprie and B. Littlewood, "Probabilistic assessment of safety-critical software: Why and how?," *Communications of the ACM*, vol. 35, no. 2, pp. 13–21, 1992.

[2] R. Yanney and J. Hayes, "Distributed recovery in fault tolerance multiprocessor networks," *4th International Conference on Distributed Computing Systems*, pp. 514–525, 1984.

[3] S. Yau and R. Cheung, "Design of self-checking software," *Proc. Int'l Conf. on Reliability Software*, pp. 450–457, April 1975.

[4] B. Randall, "System structure for software fault tolerance," *IEEE Transactions of Software Engineering*, vol. SE-1, no. 2, pp. 220–232, 1975.

[5] A. Mahmood, E. McCluskey, and D. Lu, "Concurrent fault detection using a watchdog processor and assertions," *IEEE 1983 International Test Conference*, pp. 622–628, 1983.

[6] K. Huang and J. Abraham, "Fault-tolerant algorithms and their applications to solving laplace equations," *Proceedings of the 1984 International Conference on Parallel Processing*, pp. 117–122, August, 1984.

[7] B. McMillin and L. Ni, "Reliable distributed sorting through the application-oriented fault tolerance paradigm," *IEEE Trans. of Parallel and Distributed Computing*, vol. 3, no. 4, pp. 411–420, 1992.

[8] C. Hoare, "An axiomatic basis for computer programming," *Communications of the ACM*, vol. 12, no. 10, pp. 576–583, 1969.

[9] G. Levin and D. Gries, "A proof technique for communicating sequential processes," *Acta Informatica*, vol. 15, pp. 281–302, 1981.

[10] C. Hoare, "Communicating sequential processes," *Communications of the ACM*, vol. 21, no. 8, pp. 666–677, 1978.

[11] R. Apt and W. Roever, "A proof system for communicating sequential processes," *ACM Transactions on Programming Languages and Systems*, vol. 2, no. 3, pp. 359–385, 1981.

[12] N. Soundararahan, "Axiomatic semantics of communicating sequential processes," *ACM Transactions on Programming Languages and Systems*, vol. 6, no. 6, pp. 647–662, 1984.

[13] H. Lutfiyya and B. McMillin, "Comparison of three axiomatic proof systems," *UMR Department of Computer Science Technical Report CSC91-13*, 1991.

[14] H. Lutfiyya, M. Schollmeyer, and B. McMillin, "Formal generation of executable assertions for application-oriented fault tolerance," *UMR Department of Computer Science Technical Report Number CSC 92-15*, 1992.

[15] H. Lutfiyya, M. Schollmeyer, and B. McMillin, "Fault-tolerant distributed database lock managers," *UMR Department of Computer Science Technical Report Number CSC 92-05*, 1992.

[16] H. Lutfiyya, A. Sun, and B. McMillin, "A fault tolerant branch and bound algorithm derived from program verification," *IEEE Computers Software and Applications Conference(COMPSAC)*, pp. 182–187, 1992.

[17] K. Batcher, "Sorting networks and their applications," *Proc. of the 1968 Spring Joint Computer Conference*, vol. 32, pp. 307–314, 1968.

[18] M. Quinn, *Designing Efficient Algorithms for Parallel Programs*. New York: McGraw-Hill, 1987.

[19] H. Lutfiyya and B. McMillin, "Formal generation of executable assertions for a fault-tolerant bitonic sort," *UMR Department of Computer Science Technical Report CSC91-12*, 1991.

Responsive Protocols

Towards a Responsive Network Protocol

A. SHIONOZAKI

Department of Computer Science

Keio University

3-14-1 Hiyoshi, Kohoku-ku, Yokohama 233, Japan

M. TOKORO

Sony Computer Science Labolatory, Japan

Abstract

A responsive network architecture is essential to future distributed systems. In this paper, a new model that incorporates the foundations for a responsive network architecture is proposed. This model guarantees notification of requests transmitted to the network in bounded time. Its functionalities augment existing resource reservation protocols that support multimedia communication. An overview of a real-time network protocol that is based on this model is also presented.

Key Words: Network architecture, responsive protocols, open distributed systems, real-time operating systems.

1 Introduction

Advancements in hardware technologies will make it possible for anyone to access open distributed systems regardless of time and location. For example, one will be able to communicate through packet radio networks or fast gigabit networks via portable personal computers the size of a small pocket calculator. The functionalities provided by a network architecture in such future distributed systems will be significant, because of higher user demands and greater technological advances. One such functionality is real-time communication. There will be great demands for teleconferencing software that can correctly process multimedia data

[1]. A user may want to send messages to robots that function in real-time to accomplish daily chores, or the robots may even communicate automatically among themselves. Consequently, the support of real-time communication in a network architecture is vital and is a new challenge in the field of network protocols.

Unlike a dedicated system, which is closed in itself, open distributed systems are ever changing. In a dedicated system, resources can be statically allocated to a constant number of tasks that perform computation foreseen when a system is designed. However, the indeterministic nature of dynamically changing systems presents the need for some mechanism that guarantees the availability of the required resources to complete computation. Specifically, in a network architecture, all necessary resources (processing power, memory buffers, physical network bandwidth etc.) must be reserved at each network node through which data will be transferred for a given communication session.

A reservation scheme is essential for supporting real-time communication. However, in order to effectively provide a responsive system [2], simple resource reservation procedures are insufficient, thus network architectures need to be reevaluated. In this paper, the Virtually Separated Link (VSL) model is proposed. A responsive network architecture can be constructed based on this model. This model can provide the basis for a responsive network protocol by effectively providing notification of the results of any request sent out into the network in bounded time, whether it be an acknowledgement or a failure reply and distinguishing network failures. Furthermore, through this model an application can select a recovery mechanism when needed. Finally, the design of a connection oriented network protocol based on this model called RtP (Real-time network Protocol), in which the reservation procedures and connection establishment schemes are outlined, is presented. RtP does not specify a fault model but because it is based on the VSL model it guarantees applications a way to recover from failures.

In the next section, we present the motivation behind our work. In Section 3, we introduce the VSL model, and in Section 4, we present the real-time network protocol RtP and the details of various connection establishment algorithms involved. An overview of the implementation and the current status of RtP are given in Section 5. Finally, conclusions are present in Section 6.

2 Motivation

Many workstations now have the capability to provide a real-time computing environment, such as displaying video images while simultaneously processing audio data. As a result, some effort has been made to construct multimedia programs as UNIX applications. However, consider the situation depicted in Figure 1. Data stored in a multimedia data server is transferred across two networks to be displayed at a node with a video board. Despite the availability of network media such as Fiber Distributed Data Interface (FDDI) [3], that can assure bounded response time, most network protocols that are currently in major use, such as TCP/IP and OSI do not support real-time communication in anyway whatsoever. Simply integrating them with a real-time operating system is insufficient to provide real-time communication facilities. Consequently, several panacea have been proposed that have strived to remedy this situation.

There have mainly been two approaches that have been taken to support real-time in network protocols. Protocols such as XTP (Xpress Transfer Protocol) [4], developed by Protocol Engines Inc., and VMTP (Versatile Message Transaction Protocol) [5] developed for the V-kernel at Stanford, aim to reduce protocol processing overhead and increase throughput. Both protocols incorporate a message priority scheme to optimize data transfer, but end-to-end communication delays cannot be bounded.

The other approach is to reserve necessary communication resources prior to actual data transfer, so that deadlines will not be missed. These include

- Session Reservation Protocol (SRP) [6] of DASH developed at Berkeley,

- BBN's ST-II [7], and

- Capacity Based SRP (CBSRP) [8] developed at Carnegie Mellon.

These systems support the reservation of system resources through streams or sessions in the network prior to the actual transmission of data. We believe reservation a priori to communication is the key to real-time support in a dynamically changing system, because of its inherent indeterministic nature, which makes it impossible to predetermine a precise time line. In general, these protocols reserve resources across a network for communication in one direction in the following way:

1. Reserve the network resources for output on the source node.

2. At any intermediate node, reserve the CPU for protocol processing and network resources.

3. Make reservations for the input buffers at the destination node.

4. Destination sends an acknowledgement of the establishment back to the source node.

5. Source node accepts acknowledgement

6. A session or stream is established.

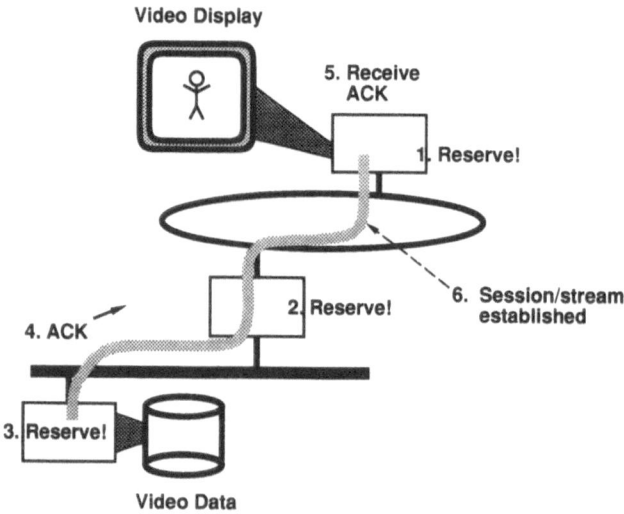

Figure 1: Reserving resources through a network.

Reservation is an effective way to provide real-time communication facilities. However, the above protocols are not responsive. When a deadline is missed or reservation requests timeout, these protocols do not provide the application the means of specifying a recovery mechanism. In particular, the reservation procedure cannot be completed in bounded time, it will result in unpredictable behavior, which is fatal for a responsive system. Also, simply relying on data links that directly support bandwidth reservation or rigid timing constraints to maintain timeliness is insufficient. A network protocol must be able to handle exceptions during communication.

Our objective in this paper is to provide the foundations for a responsive network architecture. Thus applications must be guaranteed delivery of acknowledgements or failure messages of requested services, and must be able to distinguish network failures from protocol errors, so that it can make its own decisions when it desires. Communicating tasks in responsive network architectures cannot afford to wait an indefinite amount of time for a response to arrive. A general purpose network protocol must also ensure this functionality regardless of the underlying network media.

3 The Virtually Separated Link Model

3.1 Concept

For a real-time network system to be responsive, control data such as resource reservation requests, acknowledgements, and notification of exceptions etc., which influence system execution, must to be transferred in bounded time to guarantee error handling for applications. It should be treated as expedited data with a higher priority than normal data flow, so it will not depend on current system load for data sessions. We go one step further and propose that control data should be transmitted separately from normal data. Ideally, two different physical network data links should be provided to make this possible. However, such technology is not readily at hand and is expensive, so we must compromise by encapsulating this abstraction over a single physical line at the software level.

We propose the Virtually Separated Link (VSL) model [9] in which two separate virtual channels between two communicating nodes are provided (see Figure 2). These distinguishable virtual channels are supported at the software level and provide an abstraction independent of the physical network. Thus, two different physical links are not required. This model allows a network of computational nodes to be interconnected via separate channels much like a telephone network, which provides a voice channel and a signaling channel. The VSL model provides the basic framework for real-time communication with *best effort, least suffering* properties. In other words, the system supports real-time communication for an application when it is physically possible (*best effort*), but, if it is not, the system will notify the application in bounded time, so that the application will not have to wait for an indefinite amount of time (*least suffering*). Since network behavior in an open distributed system is constantly changing and is inherently unpredictable due to failures (link failures, node failures etc.), it would be extremely expensive to provide a fault tolerance mechanism to support hard real-time. Thus, we must at least give the application a chance to recover in its own way by providing *least suffering* properties.

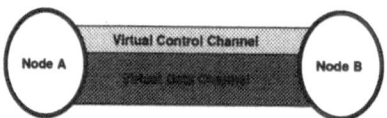

Figure 2: The VSL Model.

The channel established for exchanging control messages is called the Virtual Control Channel (VCC), and normal data is transferred through the Virtual Data Channel (VDC). The key here is that through this separation, the model provides a standard interface that emulates two different physical links. As a result, we can bound the response time of control messages for establishing data channels, reconfiguring reservation parameters, acknowledgements, and notification of failures. When a response does not arrive by the given bound the application can correctly assume that the physical link connecting its node has crashed. Control messages will always be guaranteed transmission on the VCC in the midst of normal data transfer through the VDC. In the reservation procedures of existing reservation protocols, scheduling of *open session requests* is unbounded. So, the setup time required for each session cannot be bounded, a significant drawback for a responsive network architecture.

This model is also analogous to broadband communication in which bandwidth is dedicated to one type of communication, while data for another type of communication is transferred via separate bandwidth.

3.2   Rationale

How does this separation allow us to provide a responsive network architecture? Through the establishment of the VCC, we can associate network bandwidth for control information. A protocol based on this model needs to assure only the end-to-end transfer of control messages maintaining this bandwidth limit, so all messages are processed with no overflow. Furthermore, by integrating this model with a real-time operating system, we can guarantee the schedulability of control messages carried by the VCC. The VDC and VCC also provide a standard interface for scheduling protocol processing tasks, or network media when it is possible to physically reserve bandwidth. As a result, we can transfer control messages and ensure that acknowledgements or reservation failure messages are received within a fixed time limit. In case of network failures and crashes, since we can impose a time bound, applications no longer have to wait indefinitely. In the worst case, if a response does not arrive by a specific time out value, it can choose freely to take appropriate actions. Thus either the protocol or application can choose an appropriate fault model. The VSL model guarantees the execution of error handlers to provide a predictable system.

3.3   An Example

In this section, we will illustrate how the VSL model can be used to support the notification of success or failure of communication via the VCC's within bounded time. We present an example in Figure 3.

In this example, an application on Node **A** tries to make reservations for a VDC for data transfer to its peer on Node **C**. Network message transfer times are measured by special control messages through the VCC. Processing time at each node is bounded according to a scheduling algorithm (see Section 5.1) provided by the underlying operating system.

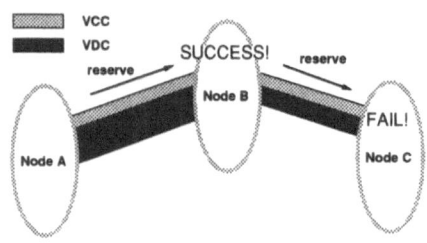

Figure 3: Example of the VSL model.

This information is collected through the VCC's and a timeout, T, is established to abort the request if it is arrives.

We will consider the case when the reservation fails at the destination node. A reservation message is sent from **A** through the VCC and the reservation succeeds at the intermediate Node **B**, but when the reservation message reaches Node **C**, it is overloaded and cannot provide any resources at the present time by T. At this point, a reserve failure message is sent back to node **A** via node **B**. If this message is received before T, **A** is notified of the overload and can decide whether or not to retransmit the request. If T arrives, **A** will know that there was a network exception between **A** and **B** or that **B** crashed, and can decide to try a different route if available. **A** knows that a network exception occurred between **A** and **B**, because in any other case it would receive a control message by T from its next hop neighbor **B**. If the link between **B** and **C** crashes, then a time out would occur at **B**, and **A** would be notified by **B**.

4   A Real-Time Network Protocol with Bounded Time Notification

4.1   Overview

In this section, we present RtP (Real-time network Protocol) [9] a network protocol, which is based on the VSL model. It does not specify a fault model, but rather guarantees that failures can be handled by applications in real-time. RtP supports the virtual channels of the VSL model and procedures that detail how these channels are established to make a virtual network of control channels interconnecting all nodes that communicate in real-time. It specifies the control messages and, through the VSL model, provides an interface to the operating system or the underlying network. RtP is a protocol intended for the network layer of the seven layer model and supports wide area networks. Presently, network level functionalities such as addressing and routing are not specified in RtP.

The functionalities provided by the underlying physical network also imposes constraints on RtP. For example, in order to completely guarantee hard real-time deadlines, a mechanism guaranteeing network capacity, as supported by FDDI or a mechanism introducing special network priorities, is necessary at the data link level. Choosing the underlying network best suited for a responsive network architecture is a question that remains to be answered.

In RtP, a channel can be established on a node basis or between session entities (applications). Channels established between nodes are called *node oriented*. Channels established between session entities are called *session oriented*. VCC's are node oriented and there is usually only one VCC between two nodes in a network. On the other hand, VDC's are session oriented, and can only be established between applications. Thus, multiple VDC's usually coexist between two nodes. Figure 4 shows a network configuration where VCC's and VDC's have been established between several nodes. In this example, requests for real-time data transfer have been made between Nodes 1 and 2, Nodes 1 and 4, Nodes 2 and 4, and, finally, Nodes 3 and 4. At this instant, there are 4 node oriented VCC's that interconnect the corresponding nodes. (The initial VCC topology depends on the type of network in question, see Section 4.3 for details. We are only considering a simple case here.) Also note that there is a one to one or more ratio between the VCC's and the VDC's as in the communication between Nodes 3 and 4. There are two VDC sessions to only one VCC.

There are three types of communication: normal datagram communication, communication via VCC's, and communication via VDC's. Normal datagram communication is basically traditional time independent communication (providing the same functionalities of protocols such as IP). The other two types involve messages transmitted via virtual chan-

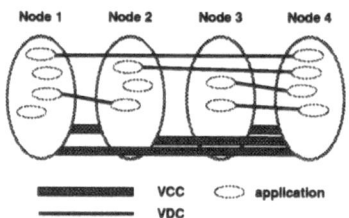

Figure 4: A network configuration example.

nels established for their particular objectives, so network transmission time can be bounded. The four ways a channel can be established in RtP are as follows.

1. Establish VCC via normal datagram communication (unbounded).

2. Establish VCC via messages through another VCC (bounded).

3. Establish VDC via normal datagram communication (unbounded, same as other reservation protocols).

4. Establish VDC via messages transmitted on a VCC (bounded).

In order to establish VDC's in bounded time, a mechanism for preallocating VCC's at system boot time is necessary. Types 2 and 4 can be used to support reservations in bounded time. The following subsections will describe in detail the procedures for channel establishment.

4.2   Establishment of VCC

To provide bounded response time for real-time communication between two nodes, a node to node VCC must be established. This VCC is referred to as the *primary* VCC between two nodes. Requests for *primary* VCC's can only be issued by using normal datagram requests or messages sent through other *primary* VCC's that have already been established. The establishment of a *primary* VCC by datagram is indeterministic, in the sense that the time required to establish it cannot be bounded. This is unavoidable in the general case, unless other VCC's have already been established to a given destination node. On subnets where there are no physical point-to-point links, *primary* VCC's must be established by datagrams (see Section 4.3). This is analogous to resource reservation protocols provided for by existing reservation protocols, except that what is being established is a VCC not a channel for actual data transfer. Note that it can be completed before any real-time channel requests from applications during system initialization. Multiple *primary* VCC's may exist between two nodes, to provide different routes. Presently, RtP does not specify any routing algorithm, but if a source routing option is available, two nodes can easily be connected by different VCC's with different routes.

Upon system initialization, a VCC_INIT control message is sent out to the nodes to which VCC's should be established (determined by mechanisms specified in Section 4.3) as depicted in Figure 5. This request must be carried via normal datagram communication. This message is forwarded to all nodes in the route to the destination node. Once a *primary* VCC is established, this step no longer needs to be taken and control messages can be transferred in bounded time.

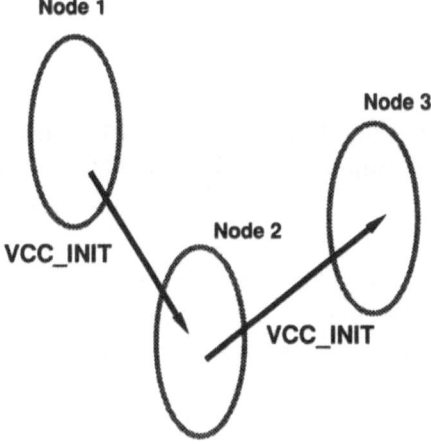

Figure 5: Primary VCC establishment request which takes place upon system initialization.

VCC's can also be established by issuing a special EST_VCC control message on an existing *primary* VCC. For certain types of communication, there might be sufficient reason to provide it with its own dedicated VCC. There are two types of EST_VCC messages: one is for establishing *secondary* VCC's to the same destination and the other is for establishing a different *primary* VCC to a different destination. This is done by issuing EST_VCC_SEC or EST_VCC_PRIM messages, respectively. A VCC established by the former method is called a *secondary* VCC with respect to the VCC used to make the establishment [1]. One that is established with the latter message is another *primary* VCC to a different destination. This is a key feature of RtP and is exploited during the initial setup of a VCC between nodes in bounded time (see Section 4.3). *Secondary* VCC's can be either node oriented or session oriented [2].

To summarize, we support two types of VCC's. A *primary* VCC is node oriented and is used to send control messages for the VDC's that are already established (which have not been established via *secondary* VCC's) or for establishing the VDC's to a corresponding node. It can also be used to establish other VCC's. A *secondary* VCC may be established for sessions or nodes that need a dedicated control channel for increased control data processing.

4.3   Preallocation of VCC

The preallocation of VCC's is essential for assuring the transfer of control messages in bounded time in order to maintain a responsive system. For bounded time communication before establishing a channel, the nodes en route from a source node to a destination node must be connected by VCC's. If there is even a single VCC missing, the response time for establishing a VDC cannot be bounded, so precautions must be taken. Guaranteed delivery will be heavily dependent on when VCC's are established. Since it is impossible to link together every node on a wide area network with each other via virtual channels a priori to any communication, the time to establish a VCC becomes a key factor. At system initialization, a node has no way of knowing which nodes it will communicate with. If there is a VCC missing in the desired channel route, the phase necessary for establishing the very first channel to a

---

[1] Note that the *secondary* VCC's cannot be established by using normal datagrams. They must be established through *primary* VCC's, hence the name.

[2] Note that *secondary* VCC's will have the same route as its corresponding primary.

new node will be unbounded. In order to avoid such situations RtP preallocates VCC's in one of the following ways.

- A node establishes VCC's to gateways on its own subnetwork at system boot time when the network interface is initialized. Gateway to gateway connections are configured in the same way a priori by system administrators[3]. This is called *speculative reservation.*

- Taking advantage of the physical properties of data links involved VCC's can be established automatically. For example on point-to-point data links VCC's can established when the network interface is initialized.

The latter completely depends upon the underlying network so it is not general and is provided merely for the sake of optimization for the special case. On the other hand, *speculative establishment* is more general and can be applied in a typical network configuration in which local area networks are interconnected via gateways. If real-time communication between subnets takes place frequently, this can prove to be effective. The speculative establishment of VCC's to gateways should be mandatory although whether it should be enforced is site dependent. If there is any other communication between nodes that is known to occur at system boot time, VCC's should be established between them.

If none of the above is possible, there will be no VCC's between two nodes. In this case, the first session entity requesting a session to a destination node through this node will have to suffer from unbounded VCC set up time.

Let us elaborate on speculative establishment using an example. A typical network configuration is illustrated in Figure 6. In this example, a session entity on source Host **A** wants to establish a session with one on Host **B**. Speculative establishment takes place as follows (in Figure 6 it has already been completed).

1. Gateway 1 sends a VCC_INIT to Gateway 2 (or vice versa), in order to establish a *primary* VCC between the two gateways.

---

[3]This case does not assume any static parameters. Parameters for gateway reservation can be dynamically changed. It is static in the sense that gateways must be configured as gateways to provide correct semantics.

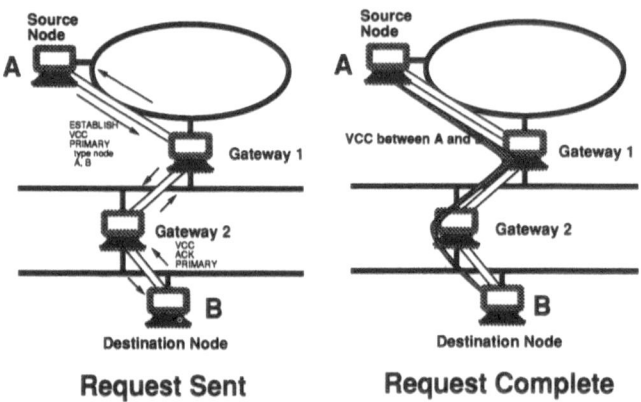

Figure 6: Speculative VCC establishment.

2. Host **A** establishes a *primary* VCC with Gateway 1, which is its local
   gateway by sending a VCC_INIT. In the same way, Host **B** establishes
   a *primary* VCC with Gateway 2.

When the session entities are ready to exchange data the following
steps are taken in order to establish a *primary* VCC between Host **A**
and Host **B** (as shown in Figure 6).

1. Host **A** sends an EST_VCC_PRIM message specifying that the VCC
   to be established is node oriented and it connects Host **A** and **B**.

2. The corresponding procedures are taken for reservation (see Section
   4.5) and, if successful, Host **B** sends out a ACK_VCC_PRIM message.

3. When the acknowledgement is returned, a *primary* VCC is established between Host **A** and Host **B**.

This procedure completes in bounded time because of the *primary*
VCC's which were established through speculative establishment during
system initialization.

4.4 Establishment of VDC

VDC's can be established in two ways as well. The first type of establishment is via normal datagram messages. This type of VDC establishment is similar to such reservation protocols as SRP and ST-II. Consequently, the establishment procedure is not completed in bounded time. It is often used in soft real-time communication in which timeliness is not critical.

Normally VDC's are established through VCC's. This type of establishment conforms with the VSL model. This is the general case, and most real-time communication will use this scheme. However, we emphasize again that before a session can establish a VDC, an appropriate VCC must be ready. When this is available, a VDC can be established in bounded time. We showed in the previous section that RtP makes special optimizations, so VCC establishment does not inhibit real-time communication.

VDC's are established through VCC's in the following way.

1. Send EST_SESSION requests via the *primary* VCC already established (see Figure 7). This is completed in bounded time. Each node that receives this request tries to make appropriate reservations for this session.

2. If Step 1 is successful, a VDC session is established for the two session entities (see Figure 8). Data transfer can now begin.

As a final note, VDC's are not node oriented. Data only needs to be transferred between two session entities.

4.5 Reservation Procedure

In this subsection, we will briefly outline the reservation procedure of RtP assuming no network failures occur. The reservation procedure for VDC's and VCC's are basically the same.

1. The requesting session entity sends a **connection request** to the RtP network server on the local node.

2. The RtP network server makes entries into system tables for the management of the virtual channel, reserves the necessary resources,

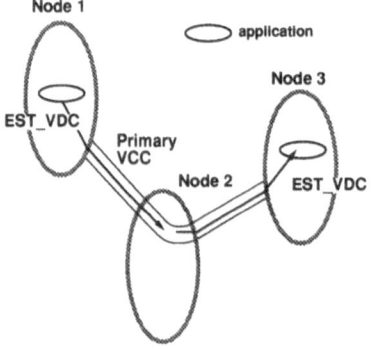

Figure 7: Establishment of VDC session between applications.

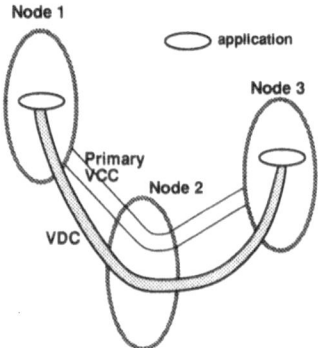

Figure 8: Establishment of VDC session completed. End-to-end real-time communication is now possible.

and, finally creates a thread or assigns an already existing thread with the appropriate execution time to the virtual channel. If this host is overloaded, then the request fails and the session entity is notified about this failure.

3. The **connection request** is forwarded to the next node en route to the node with the peer session entity via the proper communication channel (i.e., by normal datagram or VCC). This forwarding is based on a routing algorithm that is not specific to RtP. In an intermediate node, the request is sent to the RtP network server at that node. Proper network processing is completed by updating tables and reserving resources. The address for this node is also added to the *session route.*

4. If all steps are successful, the same message is forwarded to the next intermediate node and the process is repeated until the request reaches the peer node. If any step fails, a **request fail** control message is sent to the previous node of the incomplete VDC session by referring to the *session route.* **Request fail** will eventually be propagated in bounded time to the source node, then to the session entity that sent out the request. At each node, upon receiving a **request fail** message, the reserved resources are freed and the thread associated with it detached.

5. The **connection request** reaches the RtP network server at the destination node with the peer session. It makes entries into the system tables for the management of the virtual channel, reserves the necessary resources and assigns it a periodic thread. A **request acknowledge** is sent back via the intermediate nodes to the source node using the same communication channel referred to in the *session route.*

6. Each intermediate node that receives a **request acknowledge** confirms the session and forwards it back to the source node.

7. Data transmission can now commence though this virtual channel session.

5 Current Status

RtP is currently being implemented on the Mach micro kernel. We are using the x-Kernel [10] for the experimental network architecture framework. The initial implementation is being done on Ethernet to provide a testbed for media such as FDDI, which has physical network bandwidth reservation capability, and to consider the feasibility of providing a predictable soft real-time network environment for media such as packet radio. We are striving to construct a fundamental architecture that will provide us with a foundation for a general purpose implementation. As mentioned in Section 4.1, at present all network protocol functionalities have not been specified in RtP. Therefore, they must be provided separately. Presently, we augment these functions by using those of IP. Finally, we plan to change the underlying framework to Real-Time Mach [11] in the future.

5.1 Scheduling

Real-time operating systems such as Real-Time Mach [11] support various fixed scheduling policies based on the rate monotonic algorithm [12, 13] providing a suitable framework. We chose to integrate the rate monotonic algorithm with RtP over dynamic algorithms because of the following advantages. First, the period of a task can easily be mapped to a priority level depicted by integers. Since the priority of tasks can be determined with ease, it is well suited to schedule tasks that execute for network processing. It also remains stable under transient overloaded conditions [14], while for dynamic algorithms it is almost impossible to predict which task will miss a deadline under such conditions. Extensions to the rate monotonic algorithm, such as the deferred server or sporadic server, makes it possible to handle aperiodic tasks as well [15]. Any future real-time system will consist of periodic and aperiodic tasks and so it is unreasonable to leave this unconsidered. In a real-time network system, aperiodic bursts of data are not uncommon.

5.2 RtP Server

Mach's micro kernel architecture makes it possible for us to implement RtP as a user level server. As illustrated in Figure 9, an RtP network server, which is dedicated to handling all network processing, exists on each node. There are a number of high priority periodic threads that execute in its context. One of these threads, called CCH (Control Channel Handler), is a special thread that is scheduled to handle all control messages that pass through the node. It maintains and manages all VCC's associated with this node and ensures that all control information is processed before its deadline. The execution time allocated to the CCH thread is analogous to the bandwidth devoted to all the VCC's for this node when the network media does not support a network bandwidth reservation scheme of some kind. All other threads are called DCH's (Data Channel Handlers) and are dedicated to the processing of data, or rather VDC's, and are associated with the corresponding applications. An DCH is allocated for each VDC active on the node, or a specific DCH can service multiple sessions. In this way the processing time devoted to the network protocol is divided between execution time for handling control information and normal data. These time divisions must have guaranteed processor time even in the midst of other application and system programs.

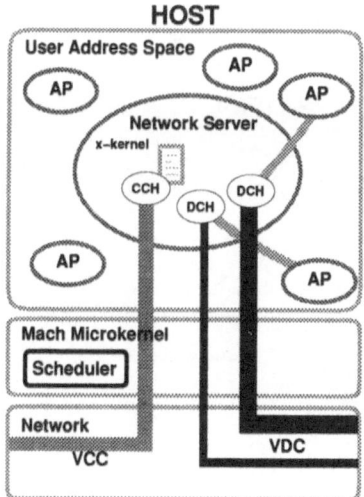

Figure 9: Design of RtP server on a micro kernel architecture.

In our current implementation, we give the CCH a period that is minimal in the system and grant it the highest priority. Consequently, the schedulability of the processing of control information can be guaranteed and a bound can be determined. This bound is essential for guaranteed notification, as shown in Section 3.3. The time required for control message flow through a VCC can be determined and this becomes the deadline for a requesting application. If an application does not receive any notification before this deadline, it can assume network failure or node crash, and take necessary actions. We are also considering the incorporation of the extended rate monotonic server algorithms such as the deferrable server or the sporadic server [15], to implement the CCH.

5.3   Application Process States

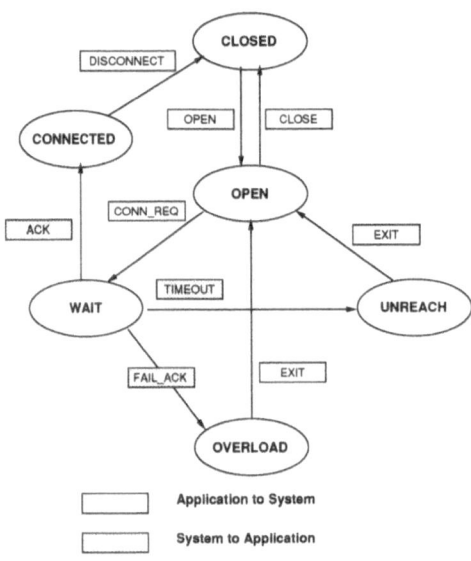

Figure 10: States of application processes (VDC establishment phase).

Part of a state transition diagram of an application during the VDC establishment phase is shown in Figure 10. The white boxes depict messages sent from the application to the CCH, and the dark boxes show events that effect its state. By collecting round trip estimate times and information of loads on other nodes, the CCH on a node is able to determine a timeout value that provides the least suffering semantics to the application. In other words, after the application goes into the WAIT state, it knows the next transition will occur in bounded time (either ACK, TIMEOUT, or FAIL_ACK). Thus, it can provide recovery mechanisms that will be guaranteed to be executed during the states OVERLOAD or UNREACH in case of network exceptions.

5.4   The VCC Network

It has been shown that dynamic reservation of resources can be effective provided there are flexible functionalities for changing reservations parameters [8]. One does not have to consider the worst case (pessimistic reservation) and can reserve resources in an optimistic way.

Thus, in our current implementation, we are focusing on optimizing VCC establishment on intermediate systems, that were doomed to be the bottleneck of VCC management. It is now not necessary for any two nodes with session entities that need to communicate in real-time, to exclusively be connected by a dedicated VCC. That is rather than viewing VCC's as point-to-point entities, VCC's are used to configure a virtual network across intermediate nodes on the internetwork. We call this interconnection the *VCC Network*.

6   Conclusion

The VSL (*Virtually Separated Link*) model was proposed to provide a foundation for responsive network protocols. Communication of control information is virtually separated at the software level from actual data transmission to ensure responsive communication for any type of underlying network medium. Applications no longer have to wait an indefinite amount of time for a response to arrive after sending out control messages or actual data into the network. The VSL framework supports the *best effort, least suffering* properties, so that a responsive protocol can be constructed on it. We are currently measuring the costs involved in incorporating the VSL model to a network protocol to justify its feasibility by implementing RtP a connection oriented real-time network protocol, and considering design issues for constructing an entire responsive network architecture.

Acknowledgements

We would like to express our gratitude to Tatsuo Nakajima and Miroslaw Malek, who provided us with valuable insight and comments on earlier drafts of this paper.

References

[1] H. M. Vin, P. T. Zellweger, D. C. Swinehart, and P. V. Rangan, "Multimedia Conferencing in the Etherphone Environment," *IEEE Computer*, vol. 24, pp. 69–79, Oct. 1991.

[2] M. Malek, "Responsive Systems: A Marriage between Real-Time and Fault Tolerance," *Fault-Tolerant Computing Systems*, vol. Informatik-Fachberichte, pp. 1–17, Sept. 1991.

[3] K. C. Sevcik and M. J. Johnson, "Cycle Time Properties of the FDDI Token Ring Protocol," *IEEE Trans. Software Engineering*, vol. SE-13, pp. 376–385, Mar. 1987.

[4] W. T. Strayer, B. J. Dempsey, and A. C. Weaver, *XTP: The Xpress Transfer Protocol.* Addison Wesley, 1992.

[5] D. R. Cheriton, *VMTP: Versatile Message Transaction Protocol Protocol Specification.* Stanford University, Feb. 1988. RFC 1045.

[6] D. P. Anderson, R. G. Herrtwich, and C. Schaefer, "SRP: A Resource Reservation Protocol for Guaranteed Performance Communication in the Internet," Tech. Rep. TR-90-006, International Computer Science Institute, Feb. 1990.

[7] C. T. et al., *Experimental Internet Stream Protocol, Version 2 (ST-II).* CIP Working Group, Oct. 1990. RFC 1190.

[8] Y. Tobe, S. T.-C. Chou, and H. Tokuda, "QOS Control for Continuous Media Communications," *Proceedings of INET '92*, pp. 471–481, June 1992.

[9] A. Shionozaki, "A Real-Time Network Protocol for Distributed Systems," Master's thesis, Keio University, Feb. 1992.

[10] N. C. Hutchinson and L. L. Peterson, "Design of the x-Kernel," in *Proceedings of the SIGCOMM '88 Symposium*, pp. 65–75, ACM, Aug. 1988.

[11] H. Tokuda, T. Nakajima, and P. Rao, "Real-Time Mach: Towards a Predictable Real-Time System," *Proceedings of USENIX 1990 Mach Workshop*, Oct. 1990.

[12] C. L. Liu and J. W. Layland, "Scheduling Algorithms for Multiprogramming in a Hard-Real-Time Environment," *J. ACM*, vol. 20, pp. 46–61, Jan. 1973.

[13] J. P. Lehoczky, L. Sha, and Y. Ding, "The Rate Monotonic Scheduling Algorithm: Exact Characterization and Average Case Behavior," 10th *IEEE Real-Time Systems Symposium*, pp. 166–171, IEEE Computer Society Press, Dec. 1989.

[14] L. Sha, J. P. Lehoczky, and R. Rajkumar, "Solutions for Some Practical Problems in Prioritized Preemptive Scheduling," 7th *IEEE Real-Time Systems Symposium*, Dec. 1986.

[15] B. Sprunt, L. Sha, and J. P. Lehoczky, "Aperiodic Task Scheduling for Hard-Real-Time Systems," *Real-Time Systems*, vol. 1, pp. 27–60, June 1989.

[16] D. Ferrari and D. C. Verma, "A Scheme for Real-Time Channel Establishment in Wide-Area Networks," *IEEE Journal on Selected Areas in Communications*, vol. 8, pp. 368–378, Apr. 1990.

[17] H. Tokuda, C. W. Mercer, and T. E. Marchok, "Towards Predictable Real-Time Communication," *Proceedings of IEEE Workshop on Real-Time Operating Systems and Software*, May 1989.

[18] J. A. Stankovic and K. Ramamritham, "What is Predictability for Real-Time Systems?," *Real-Time Systems*, vol. 2, pp. 247–254, Nov. 1990.

[19] J. Postel, *Internet Protocol – DARPA Internet Program Protocol Specification*. Defense Advanced Research Projects Agency, Sept. 1980. RFC 791.

[20] G. Agrawal, B. Chen, W. Zhao, and S. Davari, "Guaranteeing Synchronous Message Deadlines with the Timed Token Protocol," *The 12th International Conference on Distributed Computing Systems*, pp. 468–475, IEEE Computer Society Press, June 1992.

Fault-Tolerant Object by Group-to-Group Communications in Distributed Systems

H. HIGAKI

NTT

Software Laboratories

Musashino Tokyo 180, Japan

T. SONEOKA

NTT

Technology Research Department

Chiyoda Tokyo 100, Japan

Abstract

Fault-tolerance can be achieved by replicating processes. However, Precise synchronization among replicated processes, is difficult in distributed systems. This paper proposes a group-to-group communications algorithm that is an extension of the ordered multicast protocol: (1) Every replicated processes has the same execution history by letting them receive the same messages in the same order. (2) At every send event messages are sent exactly once by prohibiting backup processes from being ahead of the primary process. The algorithm does not require taking checkpoints, rollback recoveries, and broadcasting all the failure notices to all the processes. It can adopt not only server-client distributed systems but also partner model distributed systems.

Key Words: Replication, Group-to-Group Communications, Distributed Systems, Partner Model, Fault-Tolerance.

1 Introduction

With the increasing reliance being placed on distributed computer systems, high reliability and availability are becoming increasingly important. Fault-tolerance can be provided by replicating *objects* (software components) on distinct host computers.

In centralized systems, two main approaches are well known for achieving fault-tolerance using replicated objects: *passive replication approach* and *active replication approach*. Here, in order to detect failure through time-out, all the processes are assumed to be *fail-stop* [1], i.e., failed processes become silent and never send out erroneous messages to correct processes.

> **Passive replication:** Each object is replicated, but only a single replica (*the primary process*) is active and the others (*backup processes*) are passive. The primary process periodically sends the copy of its state information (*checkpoints*) to the backup processes. If the primary process fails, one of the backup processes takes it over; it rolls back to its most recent checkpoint and starts from that point.
> **Active replication:** All replicated processes are active. All processors hosting replicas are closely synchronized and execute every step simultaneously. Thus, recovery from failure of the primary process is instantaneous, because one of the backup processes can take it over without rollback recovery.

Passive replication approach has long interruption time because of rollback; restoring the state information at the most recent checkpoint and executing the event sequence already executed by the failed primary process. Furthermore, most passive replication approaches require programmers to write explicitly checkpoint settings for fault-tolerant programs.

However, in distributed systems using message passing communications (not in tightly coupled multiprocessor systems) it is difficult to achieve active replication fault-tolerance, because precise synchronization among replicas is difficult due to long message passing delays. In *server-client model* distributed systems, both approaches are realized to make only server processes fault-tolerant. But, in *partner model* distributed systems, in which all the processes are on an equal footing and communicate with each other, they have not been realized. This paper proposes a communications algorithm to achieve active replication fault-tolerance in partner model distributed systems.

Due to message passing delays, when a certain process sends messages m_a and another sends m_b to the same process group (set of replicas) simultaneously, some replicas may receive m_a first and the others may receive m_b first; the execution histories of the replicas may diverge. To overcome this problem, *ordered multicast protocols* which guarantee that all the replicas in the same process group receive the same messages (including failure notices) in exactly the same order are proposed [2, 3, 4, 5].

They are point-to-group communications protocols and can realize passive replication fault-tolerance (ISIS [5] and Auragen [6]). Some active replication approaches are proposed for server-client model distributed systems (Clouds [7]). But in partner model distributed systems,these approaches require that every replica sends the same message redundantly; when each process group consists of N times replicated processes, for a message transmission between process groups, $O(N^2)$ point-to-point (between processes) messages have to be transmitted. Even if broadcast channels are used, $O(N)$ broadcast messages have to be transmitted. Thus, too much network bandwidth are consumed and processors should manage the redundant messages.

Semi active replication approach [8] can solve this problem. At every send event only the primary process sends messages and the backup processes omit sending messages and continue to execute the successive events (Chorus system [9]). When the primary process fails, one of the backup processes takes it over. This can avoid redundant message passing; i.e., only $O(N)$ point-to-point messages ($O(1)$ broadcast messages) are transmitted.

However, omissions or duplications of sending messages might occur, when the primary process fails. On receiving the failure notice of the primary process, the backup processes cannot know which event has been completed by the failed primary process. Thus, the new primary process is not sure from which send event to start sending. As shown in Figure 1, the new primary may send messages already having been sent by the failed primary process or omit sending messages which have not been sent. The ordered broadcast protocols cannot cope with this problem, because they cannot control the ordering between sendings of messages and receptions of failure notices.

This paper proposes a group-to-group communications algorithm by extending the ordered multicast protocol for achieving semi-active replication fault-tolerance in partner model distributed systems.

The algorithm prevents omissions or duplications of message sendings even in the case of process failures by prohibiting backup processes from being ahead of the primary at every send event and by delaying the time becoming the new primary process until the forthcoming appropriate send event. Moreover, it has the following advantages:

- Interruption time by process failure is short due to no rollback

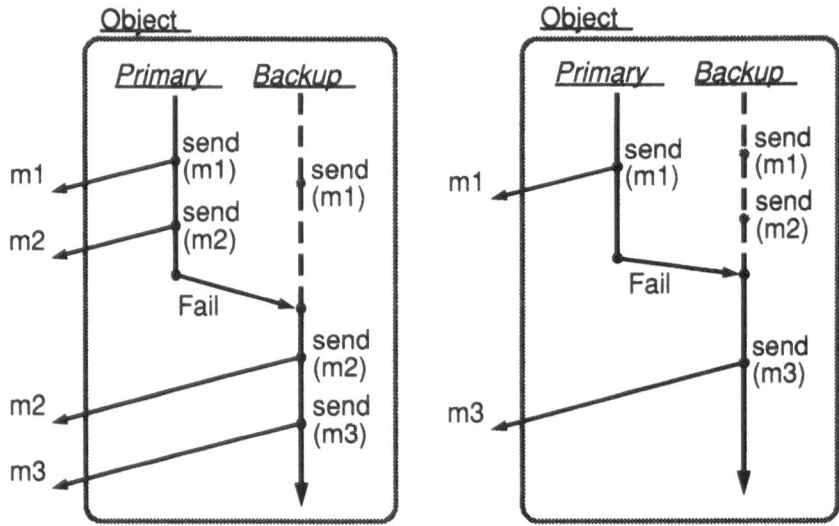

Figure 1: Duplication and Losses of Message Sendings

recovery.

• Fault-tolerance is transparent to both the programmers and the users; i.e., no additional programming such as checkpoint settings should be required.

• Failure notices are concealed in a process group.

Concerning the last point, in most systems including ISIS [5], Auragen [6], Clouds [7], processes and groups are managed in a flat and system wide name space. Every notification of process failure is broadcasted to all the processes in the system for updating their views about the process group in which the failure occurred. In large distributed systems, it would cause long interruption time for reconfiguration. The proposed algorithm conceals process failures inside a process group.

2   System Model

We consider distributed systems in which processes communicate with each other by *message passing*. Its delay is stochastically distributed, but failures are assumed to be detected by time-out. Communication channels are assumed to be FIFO (first-in-first-out) and reliable; i.e., messages are never lost, duplicated, nor contaminated.

Each process (p) is modeled by a *deterministic* state transition machine and executes three kinds of events in its application program; *receive events* (receiving message m from process q denoted by $receive_p^q(m)$), *send events* (sending message m to process q (denoted by $send_p^q(m)$), and *local events* (only computing the state of the process). A process receives a message at a receive event and executes an event sequence consisting of local events and send events which is determined by its state and the received message. Received messages are managed in the group-to-group communication layer in which the proposed algorithm is executed and will be delivered to the application program at the forthcoming receive events. Due to the determinism, if replicated processes start out form the same state and receive the same messages in the same order, their execution histories are the same and they produce the same outputs.

The type of process failures is assumed to be *fail-stop* [1]:

- Every failed process stops, becomes silent, and never sends spurious or malicious messages.
- Failures can be detected by other processes.

We assume that processes receive messages sent from a failed process before receiving its failure notice [1].

Process failures in a correct site can be detected by the monitoring function of the operating system. Site failures can be detected by transmitting '*I'm alive*' messages periodically among sites and detecting time-out [5].

Finally we assume that the following mechanisms can be used. First, processes send multiple point-to-point messages atomically; this guarantees that either these messages are sent to all the receiver processes or none of them (when the sender process fails). Second, the proposed algorithm can make messages hold the following *causal relations of messages* [10]:

Causal relation of messages: if $send_r^p(m) \rightarrow send_r^q(m')$ then $receive_p^r(m) \rightarrow receive_q^r(m')$.

---

[1] By extending the time-out threshold, the probability that the assumption holds approaches 1. Even in the case against this assumption, only duplication (no omission) might occur and can be coped with by the algorithm proposed in Appendix.

Here, the *causal* (*happens before*) relation '→' is defined on the set of events by Lamport [11] as follows:

> **Causal relation of events:** Event e_m happens before event e_n, denoted by $e_m \rightarrow e_n$, iff one of the following conditions is true:
>
> **1.** e_m and e_n occur at the same process p, and e_m precedes e_n in the local time of p.
> **2.** e_m is a send event and e_n is the corresponding receive event.
> **3.** transitive closure of 1 and 2.

Finally, the replicas of an object organize a *process group* and the proposed algorithm uses *groupview*. The groupview of G is the set of process-IDs (PIDs) in process group G and that known to process p is denoted by $\text{View}_p(G)$. For sending message m to process group G, process p sends m to each process in $\text{View}_p(G)$. Trueview(G) is defined as $\text{View}_p(G)$ of the primary process p in G, which includes all the correct processes in G.

3   Group-to-Group Communications

3.1   Requirements and Advantages

To achieve semi-active replication fault-tolerance in partner model distributed systems, the proposed algorithm should hold the following requirements in order to solve the problems described in section 1.

> **1.** At a send event only one replicated process (the primary process) sends intergroup messages exactly once. Even if the primary process fails, sendings of messages are neither omitted nor duplicated.
> **2.** All replicated processes receive the same intergroup messages in the same order. Thus, their execution histories are the same.

And the proposed algorithm has the following advantages.

> **1.** When each object is replicated N times, only $O(N)$ messages are transmitted between process groups.
> **2.** The proposed algorithm does not require checkpoint settings and rollback recoveries.
> **3.** Each process failure and recovery is informed only inside a process group immediately. Thus, global synchronization in the system is not necessary.

3.2 Principle

Avoiding Omissions and Duplications of Sending Messages

In order to guarantee that replicated processes send messages exactly once at every send event even when the primary process fails and one of the backup processes takes it over, processes should perform as follows:

> **1.** Backup processes never execute a send event before the primary process.
>
> **2.** Until the send event that the failed primary process last executed, all the replicated processes behave as backup processes at each send event.

For the former, when the primary process fails before a certain send event, all the backup processes are still before it. Thus, without using rollback recoveries, the omissions of sending messages can be avoided. To achieve this, the primary process sends intergroup messages ('Multicast' type messages) and notifications of the completion of sending intergroup messages ('Complete' type messages) atomically at each send event. For the latter, even if the backup processes perform more slowly than the primary process, intergroup messages are never sent redundantly. This is achieved by delivering failure notices ('Fail' type messages) in the causal order with 'Complete' type messages.

In this method, replicated processes do not have to be synchronized closely; the primary process does not have to synchronize with the backup processes nor wait for them at any time (backup processes may be blocked when they reach some send event before the primary process).

Delivering Messages in the Same Order

To deliver intergroup messages in the same order to the applications of the replicated processes, on receiving an intergroup message, replicated processes perform as follows: The primary process delivers the received message to its application immediately and forwards it (by 'Forward' type messages) to the backup processes. On the other hand, backup processes do not immediately deliver them to their applications and append the received intergroup messages to their message buffers. When backup processes receive the corresponding forwarded messages from the primary process, the buffered messages are delivered to their application program. Thus, backup processes deliver the received messages in the order decided by the primary process and the same order reception is available.

4   Algorithm

4.1   Preparation

Each process p has the following two message queues, one message buffer, and groupviews (Figure 2):

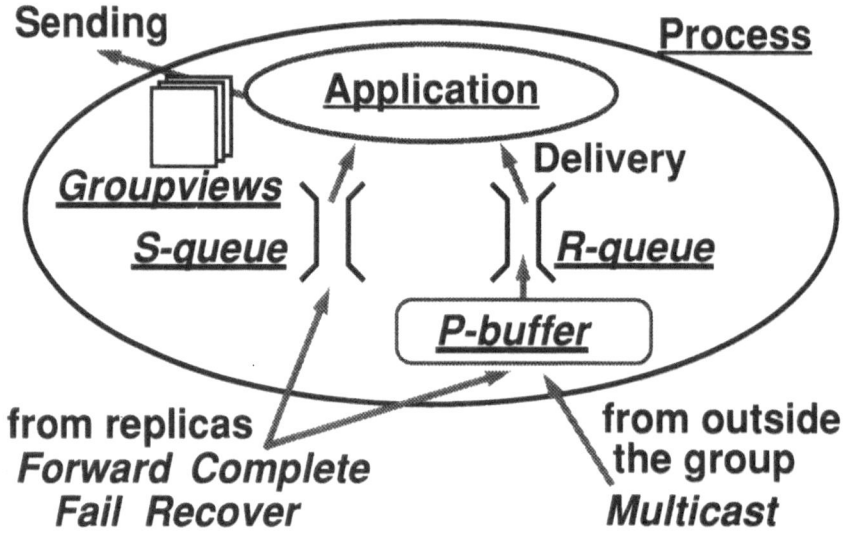

Figure 2: Process Structure

- 'S-queue$_p$' is a FIFO message queue that puts in order 'Complete', 'Fail', and 'Recover' type messages (initially empty).
- 'R-queue$_p$' is a FIFO message queue that puts in order 'Multicast' type messages (initially empty).
- 'P-buffer$_p$' is a message buffer that holds received 'Multicast' and 'Forward' type messages (initially empty).

- Groupviews $\text{View}_p(G_i)$ for every group G_i in the system (initially $\text{Trueview}(G_i)$).

Process p belonging to process group G behaves as the primary process if its PID is at the top of $\text{View}_p(G)$; otherwise, it behaves as a backup process.

Every 'Multicast' type message is assigned a unique message-ID(MID). For example, it consists of the unique group-ID of the sender and the value of the message counter indicating the number of messages sent from the process group. Furthermore, messages transmitted inside a process group ('Forward', 'Complete', 'Fail', and 'Recover' type messages) can be received in causal order using this message counter.

4.2 Message Transmissions between Process Groups

This subsection shows the algorithm for message transmissions between process groups. In the following explanation, G_s is the process group executing a send event, and G_r is the receiver process group (Figure 3).

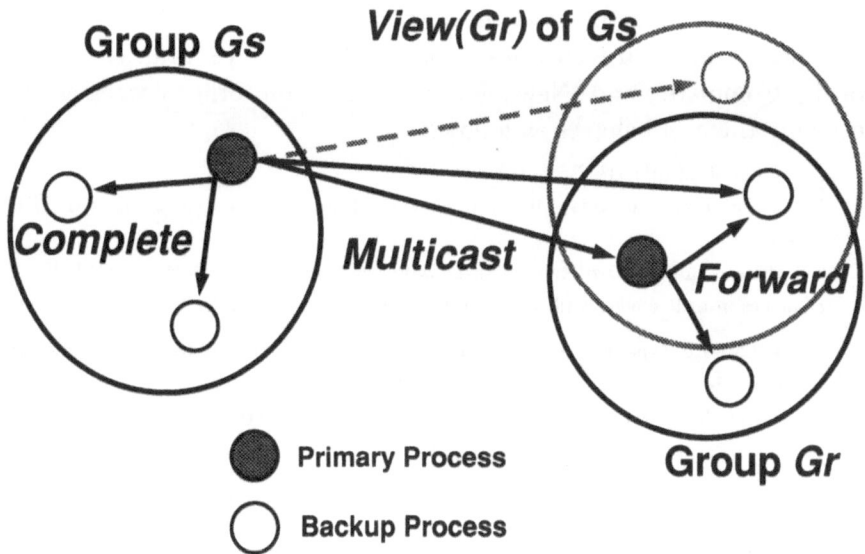

Figure 3: Overview of the Group Communication

Send event

When process p in G_s executes a send event in its application program code, p behaves as follows:

Send event (Primary) :

The primary process p sends the following two types of messages atomically.
- 'Multicast' type messages piggybacked with $\text{View}_p(G_r)$ to the processes in $\text{View}_p(G_r)$.
- 'Complete' type messages to the backup processes in G_s.

Send event (Backup) :

Backup process p dequeues a message from S-queue$_p$.
- If no message is in S-queue$_p$, p is blocked until a message is enqueued into it.
- If the message is a 'Complete' type message, p only discards it.
- If the message is a 'Fail' type message, p executes 'View renewing by failure (Backup)' in section 4.3 and restarts 'Send event'.
- If the message is a 'Recover' type message, p executes 'View renewing by recovery (Backup)' in section 4.4 and restarts 'Send event'.

Receive event

When process p in G_r executes a receive event in its application program code, p behaves as follows:

Receive event :

Process p dequeues a message from R-queue$_p$ and delivers it to its application program.

Reception of interprocess messages

Without failures and recoveries, process p can receive 'Multicast', 'Forward', 'Complete', and 'NewView' messages from other processes. On receiving them, p behaves as follows:

Reception of 'Multicast' (Primary) :

When the primary process p in G_r receives a 'Multicast' type message m with MID_m from G_s,
- if there is no 'Forward' type message with the same MID as MID_m in P-buffer$_p$, p enqueues m to R-queue$_p$ and sends the following messages atomically:

  - 'Forward' type messages with MID_m containing the copy of m to the backup processes in G_r.
  - 'NewView' type messages piggybacked with $\text{View}_p(G_r)$ to the processes in $\text{View}_p(G_s)$ if the groupview piggybacked with m is different from $\text{View}_p(G_r)$.

- otherwise, p deletes the 'Forward' type message with MID_m from P-buffer$_p$.

Reception of 'Multicast' (Backup) :

When the backup process p in G_r receives a 'Multicast' type message m with MID_m from G_s,
- if a 'Forward' type message with MID_m is in P-buffer$_p$, p deletes the 'Forward' type

message and discards m.

- otherwise, p appends m to P-buffer$_p$.

Reception of 'Complete' (Backup) :
On receiving a 'Complete' type message from the primary process, backup process p enqueues it to S-queue$_p$.

Reception of 'Forward' (Backup) :
On receiving a 'Forward' type message m from the primary process, backup process p enqueues m to R-queue$_p$. If PID$_p$ is included in the groupview piggybacked with m, and
- if a 'Multicast' type message with MID$_m$ is in P-buffer$_p$, p deletes the 'Multicast' type message.
- otherwise, p appends m to P-buffer$_p$.

Reception of 'NewView' :
On receiving a 'NewView' type message from the primary process in G_r, process p sets the groupview piggybacked with the message to View$_p(G_r)$ and treats the message as 'Multicast' type messages.

4.3   Process Failures

In this subsection the algorithm for notifications of process failures is described.

Reception of 'Fail' type messages

When process p in process group G receives a 'Fail' type message about one of the other replicated processes, p behaves as follows:

Reception of 'Fail' (Primary) :
On receiving a 'Fail' type message, the primary process p eliminates the PID of the failed process from View$_p(G)$.

Reception of 'Fail' (Backup) :
On receiving a 'Fail' type message, backup process p enqueues it to S-queue$_p$.

Renewing groupview by process failures

When backup process p executes 'Send event (Backup)' in section 4.2 and dequeues a 'Fail' type message from S-queue$_p$, p behaves as follows:

Renewing groupview by process failure (Backup) :
Process p eliminates the PID of the failed process from View$_p(G)$. If p becomes the primary process by this elimination, p executes 'Reception of 'Multicast' (Primary)' in section 4.2 for each 'Multicast' type message in P-buffer$_p$.

4.4   Process Recoveries

Informing of process recoveries

When a process is recovered in process group G, new PID is assigned to it and the new groupview is transmitted to the processes in G as follows:

Sending of 'Recover' (Primary) :

The primary process p appends the PID of the recovered process at the bottom of $View_p(G)$ and sends the following messages atomically:

• 'Recover' type messages piggybacked with PID of the recovered process to the backup processes in G.

• 'Recover' type message containing the state information and the contents of the queues, the buffer, and the groupviews in section 4.1 to the recovered process.

Reception of 'Recover' (Backup) :

On receiving a 'Recover' type message, backup processes p enqueues it to S-queue$_p$.

Reception of 'Recover' (Recovered process) :

On receiving a 'Recover' type message, the recovered process p copies the state information and the contents of the queues, the buffer, and the groupviews included in the message.

Renewing groupview by process recovery

When backup process p executes 'Send event (Backup)' in section 4.2 and dequeues a 'Recover' type message from S-queue$_p$, p behaves as follows:

Renewing groupview by process recovery :

Process p appends the PID of the recovered process at the bottom of $View_p(G)$.

5   Correctness

In this section we prove that the proposed algorithm satisfies the two requirements in section 3.1.

1. The requirement of sending intergroup messages exactly once.

2. The requirement of receiving intergroup messages in the same order.

5.1   Exactly Once Sending

Property 5.1 *Messages transmitted inside a process group ('Complete', 'Forward', and 'Recover' type messages) are received in the same order by the replicated processes.*

Proof

Only the primary process sends these messages, and no backup does so. When they are sent from the same primary process, all backup processes clearly receive them in the same order because of the assumption of FIFO channels. Consider the case that the primary process p_o in process group G sends message m_i to the backup processes and fails, and the backup processes p_n becomes the new primary process and sends message m_j. Clearly,
$$send_{p_o}^p(m_i) \rightarrow send_{p_o}^p(Fail(p_o))(\forall p \in G).$$
Process p_n should have received the 'Fail(p_o)' message before sending m_j, because p_n becomes the primary process only after receiving 'Fail(p_o)' message. Thus,
$$receive_{p_n}^{p_o}(Fail(p_o)) \rightarrow send_{p_n}^p(m_j)(\forall p \in G).$$
By the causal relations of events,
$$send_{p_o}^p(m_i) \rightarrow send_{p_n}^p(m_j)(\forall p \in G).$$
And, by the causal relation of messages,
$$receive_p^{p_o}(m_i) \rightarrow receive_p^{p_n}(m_j)(\forall p \in G). \square$$

Definition 5.1 $PID_p \succ PID_q$ $(View_r(G))$ *is defined to denote that* PID_p *is before* PID_q *in* $View_r(G)$.

Property 5.2 *For processes p_i and p_j and process-IDs PID_{q_i} and PID_{q_j} in process group G,*
$$if \ PID_{q_i} \succ PID_{q_j}(View_{p_i}(G)) \ and \ PID_{q_i}, PID_{q_j} \in View_{p_j}(G) \ then$$
$$PID_{q_i} \succ PID_{q_j}(View_{p_j}(G)).$$

Proof

Let Initview(G) be the groupview of G at the beginning. If both PID_{q_i} and PID_{q_j} are included in Initview(G), it is clearly true because Initview$_{p_i}(G)$ and Initview$_{p_j}(G)$ are the same. If $PID_{q_i} \in$ Initview(G) and $PID_{q_j} \notin$ Initview(G) then
$$PID_{q_i} \succ PID_{q_j}(View_{p_i}(G), View_{p_j}(G)).$$
Because PID_{q_j} is the process-ID of the recovered process and has been appended at the bottom of $View_{p_i}(G)$ and $View_{p_j}(G)$.
Otherwise, $PID_{q_i} \succ PID_{q_j}(View_{p_i}(G))$ means
$$receive_{p_i}(Recover(q_i)) \rightarrow receive_{p_i}(Recover(q_j)).$$
And by **Prop.5.1**,
$$receive_{p_j}(Recover(q_i)) \rightarrow receive_{p_j}(Recover(q_j)).$$

Thus,
$$\text{PID}_{q_i} \succ \text{PID}_{q_j}(\text{View}_{p_j}(G)).\square$$

Property 5.3 *At a send event 'Multicast' type messages are certain to be sent.*

Proof
Since at a send event each replicated process either sends or receives 'Complete' type messages, one of them should send the messages. And 'Complete' and 'Multicast' type messages are sent atomically, thus it never occurs that all the replicated processes go through a send event without sending 'Multicast' type messages.

It also never occurs that all the replicated processes are backup processes forever, because 'Fail' type messages are eventually received and one of them should become the primary process by **Prop.5.2**.

Consequently, 'Multicast' type messages are sure to be sent. $\square$

Property 5.4 *At a send event 'Multicast' type messages are sent at most once.*

Proof
Consider that two processes p_o and p_n satisfying $\text{PID}_{p_o} \succ \text{PID}_{p_n}$ (Trueview(both send 'Multicast' type messages at the same send event E. Since the messages are sent by the primary process, p_o sends them before its failure and p_n receives the 'Fail' type message of p_o before sending the same 'Multicast' type message at E. Thus,
$$send^p_{p_o}(Multicast(E)) \rightarrow send^p_{p_o}(Fail(p_o))(\forall p \in G)$$
$$receive^{p_o}_{p_n}(Fail(p_o)) \rightarrow send^p_{p_n}(Multicast(E))(\forall p \in G)$$
'Multicast' and 'Complete' messages are sent atomically and by the causal relation of messages,
$$receive^{p_o}_{p_n}(Complete(E)) \rightarrow receive^{p_o}_{p_n}(Fail(p_o)).$$
By causal relations of events,
$$receive^{p_o}_{p_n}(Complete(E)) \rightarrow send_{p_n}(Multicast(E));$$
a contradiction to the algorithm that every process either receives a 'Complete' type message or sends 'Multicast' type messages at a send event. $\square$

Property 5.5 *By **Prop.5.2**, **Prop.5.3** and **Prop.5.4**, 'Multicast' type messages are sent exactly once.*

5.2   Ordered Reception

Property 5.6 *If process p sends 'Multicast' type messages to process group G, the corresponding 'Forward' type messages are sure to be received by all the backup processes in G.*

Proof

Process p sends 'Multicast' type messages to the processes in $\text{View}_p(G)$. By the assumption $|\text{View}_p(G)| > k$, where k is the maximum number of process failures in G between two successive 'NewView' type messages,
$$\text{View}_p(G) \cap \text{Trueview}(G) \neq \emptyset.$$
Thus,
$$\exists p_{top} \in \text{View}_p(G) \cap \text{Trueview}(G) \; p_{top} \succ p(\forall p \in \text{Trueview}(G) - p_{top}).$$
Thus, the 'Multicast' type message is sure to be received by the primary process p_{top}. And the corresponding 'Forward' type messages are sent to all the backup processes, because $\text{Trueview}(G)$ is the same as $\text{View}_{p_{top}}(G)$. □

Property 5.7 *'Forward' type messages are sent exactly once for each 'Multicast' type message.*

Proof

By **Prop.5.4**, 'Multicast' type messages are sent exactly once, so the same primary process in the receiver process group never sends 'Forward' type messages redundantly.

Let us consider that two processes p_o and p_n, satisfying $\text{PID}_{p_o} \succ \text{PID}_{p_n}$ ($\text{Trueview}(G)$), send the 'Forward' type messages corresponding to the same 'Multicast' type message. In this case, p_o sends 'Forward' type messages and fails, and p_n receives 'Fail' type messages of p_o, becomes the new primary process, and sends the same 'Forward' type messages:
$$send^p_{p_o}(Forward) \rightarrow send^p_{p_o}(Fail(p_o))(\forall p \in G)$$
$$receive^{p_o}_{p_n}(Fail(p_o)) \rightarrow send^p_{p_n}(Forward)(\forall p \in G).$$
By the causal relations of events and messages,
$$receive^{p_o}_{p_n}(Forward) \rightarrow send^p_{p_n}(Forward)(\forall p \in G);$$
a contradiction to the proposed algorithm that the primary process does not send the same 'Forward' type messages already received. □

Property 5.8 *By **Prop.5.1**, **Prop.5.6**, and **Prop.5.7**, 'Multicast' type messages are received in the same order by all the replicated processes.*

6   Overhead

In this section we evaluate the overhead of the proposed algorithm, where each object is replicated N times and the total number of processes in the system is M (usually $M \gg N$), and compare with ISIS system[5], one of the typical systems implementing the ordered multicast protocol.

For intergroup communications, the proposed algorithm transmits $3N - 2$ point-to-point messages, that is N 'Multicast' type messages, $N - 1$ 'Complete' type messages, and $N - 1$ 'Forward' type messages. For a notification of a process failure, $N - 1$ messages are transmitted immediately. Additionally, for renewing the groupviews $(M - N)/\bar{k}$ messages per one process failure are required on average, where $\bar{k}$ is the average number of failures between successive 'NewView' type messages. On the other hand, ISIS requires $3N$ messages for intergroup communications (*ABCAST* protocol), and $3M$ messages for a failure notice (*GBCAST* protocol).

The time overhead is evaluated by the worst delay between the send event and the corresponding receive event, where Δ is the worst communication delay between two processes. Since 'ABCAST' and 'GBCAST' in ISIS are implemented in 2-phase, the communication delay is 3Δ for both an intergroup message and a failure notice. On the other hand, the proposed algorithm requires 2Δ ('Multicast' and 'Forward' type messages) for an intergroup message and Δ ('Fail' type message) for a failure notice.

Thus, with less overhead than that of ISIS, the proposed algorithm can achieve semi-active replication fault-tolerance in partner model distributed systems.

7   Conclusions

This paper proposes a new group-to-group communications algorithm for achieving semi-active replication fault-tolerance in partner model distributed systems. The algorithm guarantees the same ordering of message receptions in a process group and no message sending omission and duplication even in the presence of process failures, by prohibiting backup processes from being ahead of the primary process. Thus, the algorithm has the following advantages:

- In spite of replication, time overhead caused by synchronization is small, because the primary process does not need wait for the backup processes.
- Interruption time by process failure is short due to no rollback recovery.
- Fault-tolerance is transparent to both the programmers and the users; i.e., no additional programming such as checkpoint settings should be required.
- Overhead to treat process failures is small because failure notices are concealed in a process group.

Appendix: Extended Algorithm

A.1 Failure Model

The algorithm in section 4 assumes that message losses never occur in communication channels and that all messages can be sent atomically. In this section, for relaxing these assumptions, an extended algorithm is proposed.

Lower layer function

For messages transmitted between processes in the same group, by using the 'acknowledgment protocol' in the lower layer, message losses are concealed from the upper layers. Since processes are informed all the failures in a process group, the primary process can eventually receive either an 'Ack' type message or a 'Fail' type message, or when it receives neither of them, it knows that the message has been lost in the communication channel. Thus, by sending the message repeatedly until the primary process receives either an 'Ack' type message or a 'Fail' type message [2], message losses can be concealed and atomic message sending is available.

On the other hand, for intergroup messages we cannot use this protocol. If process p sends a message to process group G and process q in $View_p(G)$ has already failed, p can receive no message and should repeat sending the message forever because the failure notification of q are sent

---

[2]Here, we make an assumption that 'Fail' messages are also sent by using acknowledgment protocol from the operating system having hosted the failed process.

only inside G; in the presence of losses of messages in communication channels, the atomic sending of intergroup messages is not available.

A.2   Requirements

This algorithm satisfies the following three requirements:

1. At a send event only one replicated process sends ingergroup messages. Even when the process fails while executing the algorithm to send these messages, omissions of sending messages never occur.
2. All replicated processes in a process group receive intergroup messages in the same order.
3. Each process failure and recovery is informed only inside a process group.

A.3   Overview

An overview of the extended algorithm is as follows (Figure 4):

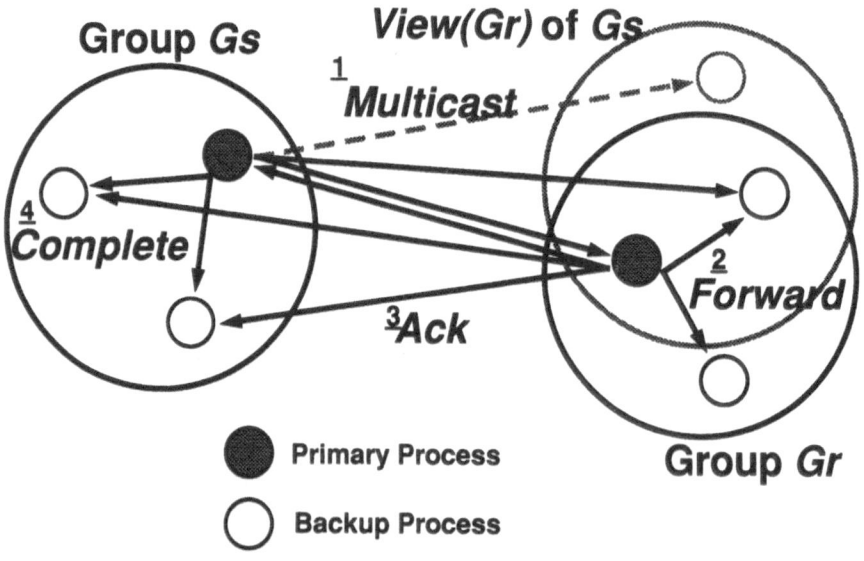

Figure 4: Overview of the Extended Algorithm

Intergroup messages

G_s and G_r are the sender and the receiver process groups, respectively.

1. The primary process p_s of G_s sends the 'Multicast' type messages to the processes in $\text{View}_{p_s}(G_r)$, sets up a time-out procedure with the threshold of $T.O.$, and waits for an 'Ack' type message from G_r. If it passes $T.O.$ without receiving an 'Ack' type message, p_s sends the same 'Multicast' again.

2. On receiving a new 'Multicast' type message, the primary process p_r of G_r sends the corresponding 'Forward' type messages to the backup processes in G_r, and sends 'Ack' type messages to the processes in $\text{View}_{p_r}(G_s)$.

3. On receiving the 'Ack' type message, p_s sends 'Complete' type messages to the backup processes in G_s.

4. The backup processes in G_s are blocked at this send event until receiving the 'Complete' type message from the primary process.

Failure notices

The algorithm for a process failure in process group G is as follows:

1. On dequeuing a 'Fail' type message from S-queue$_p$ at a send event, p eliminates the PID of the failed process.

2. If p becomes the new primary by this elimination,

2.1 if the last received message from the failed primary is a 'Forward' type message, then p sends the 'Ack' type messages corresponding to it, because the failed primary process may have received the 'Multicast' type message, sent the 'Forward' type messages, and failed without sending the 'Ack' type messages.

2.2 Process p waits for the 'Ack' type message for the send event, because the failed primary may have sent the 'Multicast' type message and failed before sending the 'Complete' type messages.

Timestamp

'Multicast' type messages may be received more than once because of losses of messages in channels, and 'Ack' type messages may be received redundantly because of failure of the primary process in the receiver process group. Not to deliver intergroup messages redundantly to the application programs, the algorithm uses timestamps, which are also used to make correspondences between 'Multicast' type messages and 'Forward' type messages or between 'Ack' type messages and 'Complete' type messages.

A.4   Preparation

Each process p has the following:

- S-queue$_p$, R-queue$_p$, P-buffer$_p$, and View$_p(G_i)$, which are the same as that in section 4.1.
- L-message$_p$ to keep the last received 'Forward' type message.
- Timestamp $TS_p(G_i)$ to manage the received intergroup messages for every G_i.

All messages are transmitted with message timestamp TS(Message) which are assigned by the sender process.

A.5   Message Transmissions between Process Groups

G_s and G_r represent the sender and the receiver groups, respectively.

Send event

At a send event in the application, p executes as follows:

Send event (Primary) :

The behavior of the primary process p consists of the following three phases:

(Multicast Phase) Process p increments $TS_p(G_s)$ and sends 'Multicast' type message, where TS(Multicast):=$TS_p(G_s)$, to the processes in View$_p(G_r)$.

(Acknowledge Phase) Process p sets up a time-out $T.O.$ and waits for the 'Ack' type message satisfying TS(Ack):=$TS_p(G_s)$. When it passes $T.O.$ without receiving it, p sends the same 'Multicast' type message and waits again.

(Complete Phase) On receiving the 'Ack' type message, p sends 'Complete' type messages, where TS(Complete):=TS(Ack), to the backup processes in G_s atomically.

Send event (Backup) :

The same as 'Send event (Backup)' in section 4.2.

Receive event

The same as 'Receive event' in section 4.2.

Reception of interprocess messages

Process p may receive 'Multicast', 'Forward', and 'Complete' messages from other processes and manages them as follow:

Reception of 'Multicast' (Primary) :

Process p enqueues the received 'Multicast' type message to R-queue$_p$, sets its local timestamp $TS_p(G_s)$:=TS(Multicast), sends 'Forward' type messages containing the copy of the 'Multicast' type message with TS(Forward):=TS(Multicast) to the backup processes in View$_p(G_r)$ atomically, and sends 'Ack' type messages with TS(Ack):=TS(Multicast) to the processes in View$_p(G_s)$.

Reception of 'Multicast' (Backup process) :
- If TS(Multicast)>$TS_p(G_s)$, process p appends the received 'Multicast' type message to P-buffer$_p$ and sets $TS_p(G_s)$:=TS(Multicast).
- Otherwise, p discards it.

Reception of 'Complete' (Backup) :
Process p sets L-message$_p$:=$\emptyset$ and enqueues the received 'Complete' type message to S-queue$_p$.

Reception of 'Forward' (Backup) :
Process p appends the received 'Forward' type message to R-queue$_p$, sets the message to L-message$_p$, and deletes the 'Multicast' type messages, which are sent from G_s and satisfy the condition TS(Multicast)$\leq$TS(Forward), from P-buffer$_p$. If TS(Forward)>$TS_p(G_s)$, p sets $TS_p(G_s)$:=TS(Forward).

A.6   Process Failure

The algorithm for 'Fail' type messages in process group G is as follows:
Reception of 'Fail' messages
The same as 'Reception of 'Fail'' in section 4.3.
Renewing groupview by process failure
If backup process p dequeues a 'Fail' type message from S-queue$_p$ at a send event, p executes the following:

Renewing groupview by process failure (Backup):
Process p eliminates the PID of the failed process from View$_p(G)$. If p becomes the primary process by this elimination, p executes the following:
- If a 'Forward' type message has been set to L-message$_p$, p sends 'Ack' type messages with TS(Ack):=TS(Forward) to the processes in View$_p(G_s)$, where G_s is the sender process group of the corresponding 'Multicast' type message.
- Each 'Multicast' type message m in P-buffer$_p$ is treated as follows:
⋆ Process p sets $TS_p(G_s)$:=TS(m), where G_s is the sender process group of m, enqueues m to R-queue$_p$, and deletes m from P-buffer$_p$.
⋆ Process p sends 'Forward' type messages containing the copy of m to all the backup processes in G and 'Ack' type messages with TS(Ack):=TS(m) to the processes in View$_p(G_s)$.
- Process p increments $TS_p(G)$ and starts 'Send event (Primary)' in section A.5 from (Acknowledge Phase).

A.7   Process Recovery

The algorithm for process recovery is the same as that in section 4.4

Acknowledgement

The authors greatly appreciate the encouragement and the suggestions given by Dr. H.Ichikawa, Dr. Y.Hirakawa in NTT Software Laboratories, and Mr. Y.Manabe in NTT Basic Research Laboratories.

References

[1] F. Schneider, "Byzantine generals in action: implementing fail-stop processors," *ACM Trans. on Comp. Syst. 2(2)*, pp.145 – 154 (1984).

[2] J. Chang and N.F. Maxemchuk, "Reliable broadcast protocols," *ACM Trans. on Comp. Syst. 2(3)*, pp.251 – 273 (1984).

[3] F. Cristian, H. Aghili, and R. Strong, "Atomic Broadcast: From Simple Message Diffusion to Byzantine Agreement," *Proc. of the 15th FTCS*, pp.200 – 206 (1985).

[4] M.F. Kaashoek and A.S. Tanenbaum, "Group communication in the Amoeba distributed operating system," *Proc. of the 11th ICDCS*, pp. 222 – 230 (1991).

[5] K. Birman and T. Joseph, "Reliable communications in presence of failures,"

ACM Trans. on Comp. Syst. 5(1), pp. 47 – 76 (1987).

[6] A. Borg, J. Baumbach, and S. Glazer, "A message system supporting fault tolerance," Proc. of the 9th ACM Symp. on OS Principles, pp.90 – 99 (1983).

[7] M. Ahamad, P. Dasgupta, R. LeBlanc, and C.T. Wilkes, "Fault tolerant computing in object based distributed operating systems," *Proc. of the 6th Symp. on Reliable Distributed Systems*, pp. 115 – 125 (1987).

[8] P.A. Barrett, A.M. Hilborne, P.G. Bond, D.T. Seaton, P. Verissimo, L. Rodrgues, and N.A. Speirs, "The Delta-4 extra performance architecture (XPA)," *Proc. of the 20th FTCS*, pp. 481 – 488 (1990).

[9] J.S. Banino and J.C. Fabre, "Distributed coupled actors: a Chorus proposal for reliability," *Proc. of the 3rd ICDCS*, pp. 128 – 134 (1982).

[10] A. Schiper, J. Eggli, and A. Sandoz, "A new algorithm to implement causal ordering," *Lecture Notes in Computer Science 392, Distributed Algorithms*, pp.219 – 232 (1989).

[11] L. Lamport, "Time, clocks, and the ordering of events in a distributed system," *Commun. of the ACM 21(7)*, pp. 558 – 565 (1978).

Space-Time Tradeoff in Hierarchical Routing Schemes

K. ISHIDA*

Department of Information Science

Hiroshima Prefectural University

Shobara-shi, Hiroshima 727, Japan

Abstract

A foundation of responsive system design lies in consideration of space-time tradeoff. Routing a message in computer networks is efficient when each node knows the full topology of the whole network. However, in the hierarchical routing schemes, no node knows the full topology.

In this paper, a tradeoff between an optimality of path length (message delay: time) and the amount of topology information (routing table size: space) in each node is presented. The schemes to be analyzed include *K-scheme* (by Kamoun and Kleinrock), *G-scheme* (by Garcia and Shacham), and *I-scheme* (by authors). The analysis is performed by simulation experiments. The results show that, with respect to average path length, *I-scheme* is superior to both *K-scheme* and *G-scheme*, and that *K-scheme* is better than *G-scheme*. Additionally, an average path length in *I-scheme* is about 20% longer than the optimal path length. For the routing table size, three schemes are ranked in reverse direction. However, with respect to the order of size of routing table, the schemes have the same complexity $O(\log n)$ where n is the number of nodes in a network.

Key Words: Large computer network, hierarchical routing, routing table size, optimality of path length.

1 Introduction

Computer networks have been increasing in size rapidly and come to contain more than order of hundreds of nodes. Among several network

---

*Part of this work was performed while the author was staying at University of Texas at Austin.

control functions, routing is an essential one, and it is needed to be fault tolerant and real-time. Routing a message in computer networks is efficient when each node knows the full topology of the whole network. This may not be the case in large networks because the network may be composed of smaller autonomous regions by design or by requirements on performance. Each region has less than complete information about other regions. For the large networks, hierarchical routing schemes have been proposed [1, 2, 3, 4]. In hierarchical routing schemes, each node has less than complete information about other regions.

To develop a fault-tolerant real-time (called responsive) system, it is needed that ingenious ways of adding some redundancy (extra space or extra time). In other words, a foundation of responsive system design lies in consideration of space-time tradeoff [5]. However, insufficient investigation is performed for space-time tradeoff in hierarchical routing schemes.

In this paper, a tradeoff between an optimality of path length (message delay: time) and the amount of topology information (routing table size: space) in each node is presented. The schemes to be analyzed include *K-scheme* (by Kamoun and Kleinrock) [3], *G-scheme* (by Garcia and Shacham) [1], and *I-scheme* (by authors) [2, 6]. The analysis is performed by simulation experiments. The results show that, with respect to average path length, *I-scheme* is superior to both *K-scheme* and *G-scheme*, and that *K-scheme* is better than *G-scheme*. Additionally, an average path length in *I-scheme* is about 20% longer than the optimal path length. For the routing table size, three schemes are ranked in reverse direction. However, with respect to the order of size of routing table, the schemes have the same complexity $O(\log n)$ where n is the number of nodes in a network.

2   Network Model

2.1   Terminology

A computer network is modeled by a connected and undirected graph $G=(V, E)$, called network graph. A node $v_i \in V$ and an edge $(v_j, v_k) \in E$ correspond to a node and a communication line between v_j and v_k in the computer network, respectively. Each undirected edge of G is labeled

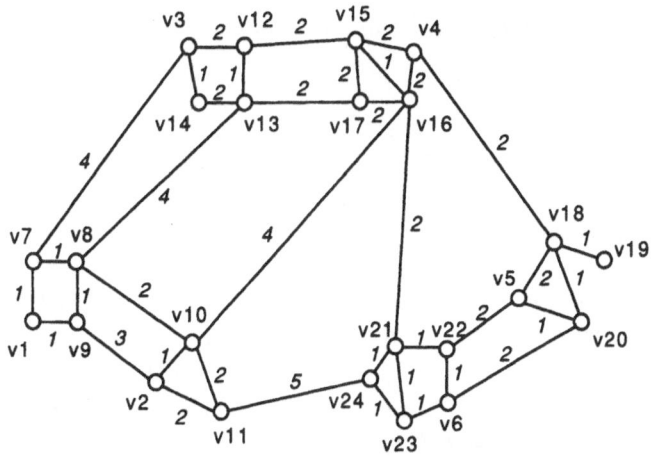

Figure 1: An example of network graph.

by a weight (≥ 0), which corresponds to a communication delay for the edge. The label is called a weight of the edge.

Definition 1 *A path on $G = (V, E)$ is defined as a sequence of nodes $v_{i_1}, v_{i_2}, \ldots, v_{i_t}$ such that all nodes are distinct and $(v_{i_j}, v_{i_{j+1}}) \in E$, $1 \leq j < t$. The length of a path P, denoted by $|P|$, is defined as the sum of the weights of the edges along the path P.*

Definition 2 *The distance between v_j and v_k on a network G is defined as the minimum length among all paths between v_j and v_k on G.*

Definition 3 *For a network graph $G=(V, E)$ and a set of nodes $C = \{c_1, c_2, \ldots, c_m\}$ $(c_i \in V)$, a clustering $CL(G)$ is defined as a partition $\{V_1, V_2, \ldots, V_m\}$ of the set V such that each subgraph $G_i = (V_i, E_i)$ of G induced by V_i (that is, $V_i \subseteq V$, $E_i = (V_i \times V_i) \cap E$) is a connected graph, and $V_i \cap C = \{c_i\}$. Each c_i is called a control node, and each $G_i = (V_i, E_i)$ is called a cluster.*

A clustering of a network constitutes a clustered network that consists of a set of clusters. In the clustered network, each cluster is regarded as an independent subnetwork that is managed by a control node [2, 7]. The network management can be carried out in a distributed manner.

2.2   Cluster-Based Network

Definition 4 *A k-level clustering and a k-level cluster are defined as (1) and (2) recursively.*

(1) *Consider a clustering $CL(G) = \{G_1, G_2, \ldots, G_m\}$. For each $G_i = (V_i, E_i)$ $(1 \leq i \leq m)$, $|V_i|=1$ and $|E_i|=0$ hold, $CL(G)$ is defined as 0-level clustering and represented by $CL^{(0)}(G)$. Each G_i is called 0-level cluster and represented by G_i^0.*

(2) *Assume that $CL^k(G) = \{G_1^k, G_2^k, \ldots, G_s^k\}$, $G_i^k = (V_i', E_i')$ $(1 \leq i \leq s)$, is k-level clustering. Then, $CL(G) = \{G_1", G_2", \ldots, G_t"\}$ with $G_j" = (V_j", E_j")$ $(1 \leq j \leq t)$ satisfying the following C.1–C.3, is defined as (k+1)-level clustering $CL^{(k+1)}(G)$, and each $G_j"$ is defined as (k+1)-level cluster G_j^{k+1}.*

C.1 For $1 \leq i \leq s$ and $1 \leq j \leq t$, $V_i' \cap V_j" = V_i'$ or ϕ.
C.2 $1 \leq t \leq s - 1$
C.3 $G_j"(1 \leq j \leq t)$ is connected.

From Definition 4, a whole network G is regarded as a cluster. Assume that the highest level of clusterings in G with $t \geq 2$ is h-1 $(h \geq 2)$ level, and then G is considered as an h-level cluster. Then, the network G is regarded as an h-level hierarchical network, denoted by G^h.

Definition 5 *For a k-level $(k \geq 1)$ cluster $G_i = (V_i, E_i)$, a set of k-level border nodes B_i for G_i is defined as follows:*

$$B_i = \{ v_j | (v_j, v_x) \in E \wedge v_j \in V_i \wedge v_x \notin V_i \}.$$

A k-level border node in a k-level cluster sends (or receives) messages to (or from) other k-level border nodes in different k-level clusters.

Example 1 *Figure 1 shows an example of network graph $G = (V, E)$, $|V|=24$. Figure 2 shows an example of 3-level hierarchical network G^3 for the network graph G. Each node in Figure 2 is identified by a cluster name followed by a node number, e.g., A.1.4. Six control nodes $c_1 = $ A.1.4, $c_2 = $ A.2.3, $c_3 = $ B.1.1, $c_4 = $ B.2.2, $c_5 = $ C.1.4 and $c_6 = $ C.2.3 are assumed. In the cluster A, nodes A.1.1, A.1.2, A.2.1 and A.2.2 are 2-level and 1-level border nodes. A.1.3 and A.2.3 are 1-level border nodes. Each label attached to an edge denotes the weight.*

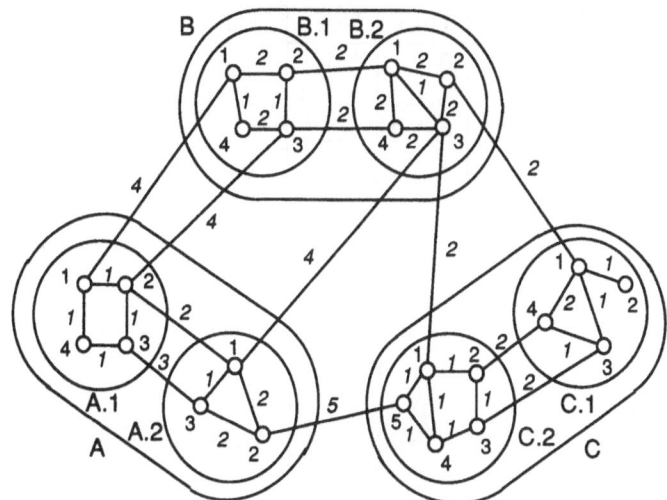

Figure 2: An example of 3-level hierarchical network G^3.

3 Hierarchical Routing Schemes

3.1 Hierarchical Routing Table

In a routing scheme, hierarchical routing tables, stored at a plain node (i.e., the node except for control node and border node), can be divided semantically into two distinct routing tables: LRT (Local Routing Table) and GRT (Global Routing Table). A node in a cluster G_i^1 refers to its LRT(or GRT) when the node sends a message to any node inside (or outside) the same 1-level cluster G_i^1.

According to information included in GRT at a plain node, most schemes can be classified into the following three types.

(1) For each cluster, the shortest path from the node to exactly one border node is stored in the routing table [3, 4, 8]. Kamoun's scheme [3] is a representative of this type. Then, this type is referred to as *K-scheme.*

(2) This is the same as (1) except that the definition of a distance between two 1-level clusters is defined as the number of 1-level clusters

Dest	Next	Dist
A.1.1	A.1.1	1
A.1.2	A.1.3	2
A.1.3	A.1.3	1
A.1.4	---	0

(a) LRT.

Dest	Next	Dist
A.1	---	0
A.2	A.1.3	4
A	---	0
B	A.1.1	5
C	A.1.3	10

(b) GRT.

Figure 3: Routing tables at node $A.1.4$ in K-scheme.

along the path. Garcia's scheme [1] is a typical scheme of this type. Then, we call this type *G-scheme*.

(3)   For each cluster, the multiple shortest paths from the node to more than one border node are stored in the routing table. The routing scheme proposed by the authors [2] has the shortest paths from the node to all border nodes for each cluster. This type is referred to as *I-scheme*.

3.2   Kamoun's Routing Scheme

Figure 3 shows LRT and GRT in *K-scheme* at node A.1.4 in Figure 2. LRT contains an entry for each node in the same 1-level cluster. GRT contains an entry for each cluster in the k-level $(k \geq 2)$ cluster containing the node.

Since LRT is similar to a conventional routing table [3], we only explain the interpretation of GRT in Figure 3. The second row (A.2, A.1.3, 4) in the GRT implies that if node A.1.4 sends a message to any node in the cluster A.2, the next node is A.1.3, and the length of the path is 4 from A.1.4 to a nearest border node of cluster A.2.

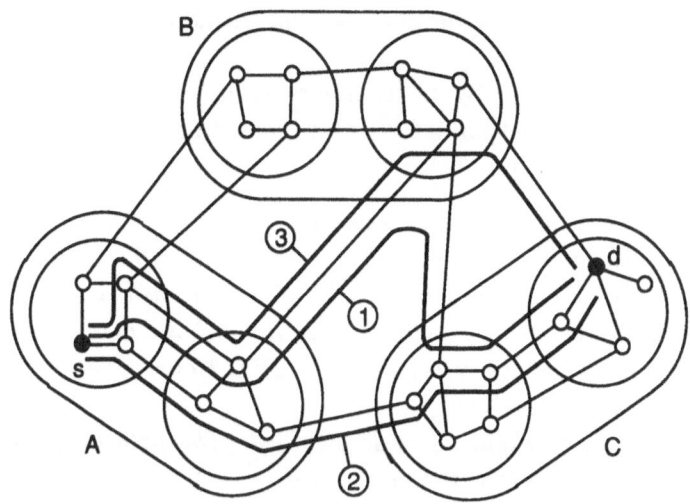

Figure 4: ① K-path, ② G-path, and ③ I-path from s to d.

Example 2 *Let us consider a routing from source node s $=A.1.4$ to destination node d $=C.1.1$ in Figure 2. The node s recognizes by referring to d's address $C.1.1$ that d belongs to a different cluster. Then, s determines the next node $A.1.3$ by referring to GRT in s. Consequently s transmits a message to $A.1.3$.*
As a result, a path ① in Figure 4 is determined. In this case, the path length is 15 and the path is not a shortest path from s to d.

The path length obtained by *K-scheme* in the worst case is much longer than the shortest path [1, 2]. This originates from the degree of utilization of routing information. The *K-scheme* uses exactly one border node for each cluster to determine a next node, even though there are multiple border nodes for each cluster in general.

3.3 Garcia's Routing Scheme

Figure 5 shows LRT and GRT at node A.1.4 in Figure 2 of *G-scheme*. Note that LRT is the same as the one used in *K-scheme*. In GRT, BN (Boundary Node) and GRN (Group Routing Node) are introduced into a hierarchical network [1]. BN is defined as the node that is connected with nodes in different 1-level cluster. Border node in Definition 5 is an extension of BN, because border node has a hierarchical level but BN doesn't.

A message transmission by *G-scheme* is explained in the following example. It is assumed that GRNs in Figure 2 are A.1.3, A.2.3, B.1.2, B.2.1, C.1.2, and C.2.4.

Example 3 *Consider the same s (= A.1.4) and d (= C.1.1) as those in Example 2. The node s recognizes that C.1.1 is contained in a different cluster C. Then, s obtains next GRN A.2.3 toward d by using GRT in Figure 5. Successively s obtains next BN A.1.3. Then, s determines A.1.3 as the next node by referring to its LRT. Consequently, s transmits to A.1.3 a message destined for d.*

Thus, a path ② in Figure 4 is determined. The path length is 17 and the path isn't a shortest path from s to d.

G-scheme has the same defect of *K-scheme*, which is mentioned in Subsection 3.2. Moreover, the path length in the worst case obtained by *G-scheme* is longer than the path length by *K-scheme* [2]. This is derived from the fact that the path length of *G-scheme* is estimated by the number of 1-level clusters.

3.4 I-Scheme Based on Border Nodes

We have already proposed *I-scheme* to overcome the weaknesses of *K-scheme* and *G-scheme* with respect to the optimality of path length [2, 6]. In *I-scheme*, any node i in a hierarchical network G^2 can send a message to any border node along a shortest path. The detailed characteristics are explained in the following example.

Example 4 *A node A.1.4, for example, in Figure 2 can send a message to each of 1-level border nodes A.2.1, A.2.2, and A.2.3 within cluster A.2 along the shortest paths on cluster A. Additionally, A.1.4 can also send a message to each of 2-level border nodes B.1.1, B.1.3, B.2.2, B.2.3, C.1.1, C.2.1, and C.2.5 along the shortest paths on the whole network G^3.*

In *I-scheme*, each node contains routing tables LRT and GRT as shown in Figure 6. Note that LRT is the same as the one used in *K-scheme*. We explain a routing based on *I-scheme* by using an example.

Dest	Next	Dist
A.1.1	A.1.1	1
A.1.2	A.1.3	2
A.1.3	A.1.3	1
A.1.4	—	0

(a) LRT.

Dest	Next GRN	Next BN	Dist
A.1	—	—	0
A.2	A.2.3	A.1.3	1
A	—	—	0
B	B.1.2	A.1.1	1
C	A.2.3	A.1.3	2

(b) GRT.

Figure 5: Routing tables at node $A.1.4$ in *G-scheme*.

Example 5 *Let us consider the same s (=A.1.4) and d (=C.1.1) as those in Example 2. The node s recognizes that C.1.1 belongs to a different cluster C.*
Then, by using the GRT, s determines destination border node C.1.1 and next border node A.1.3. To transmit a message to A.1.3, s looks up next node in the LRT. In this example, next node is A.1.3 itself. Consequently, s transmits a message to A.1.3.
Thus, a path ③ in Figure 4 is determined. The path length is 12 and the path is a shortest path from s to d on the network G^3.
If a destination node d is not a border node, a source node s sends a message to a nearest border node in a cluster to which d belongs. For example, assume that s=A.1.4 and d=C.2.4. Then s transmits a message to a nearest border node C.2.1 via next border node A.1.3.

We have already reported the theoretical results of comparing the path lengths obtained by *K-scheme*, *G-scheme*, and *I-scheme* (see Table 1) [2, 6]. The results are explained intuitively in the following. The worst case occurs in both *K-scheme* and *G-scheme* when the destination node of the path is a border node. As mentioned in Example 4, any node in *I-scheme* can send a message to the border node along a shortest path.

Dest	Next	Dist
A.1.1	A.1.1	1
A.1.2	A.1.3	2
A.1.3	A.1.3	1
A.1.4	—	0

(a) LRT.

Dest	Dest BdrN	Next BdrN	Dist
A.1	—	—	0
A.2	A.2.1	A.1.3	4
A.2	A.2.2	A.1.3	6
A.2	A.2.3	A.1.3	4
A	—	—	0
B	B.1.1	A.1.1	5
B	B.1.3	A.1.1	6
B	B.2.2	A.1.3	10
B	B.2.3	A.1.3	8
C	C.1.1	A.1.3	12
C	C.2.1	A.1.3	10
C	C.2.5	A.1.3	11

(b) GRT.

Figure 6: Routing tables at node $A.1.4$ in *I-scheme*.

Therefore, *I-scheme* is superior to both *K-scheme* and *G-scheme* with respect to path length in the worst case.

4   Tradeoff between Path Length and Routing Table Size

Simulation study is appropriate to evaluate performance of routing schemes in an average case. The three schemes are compared with respect to optimality of path length and routing table size. The simulation was carried out by using the C language on Sun Workstation.

Definition 6 *For a source-destination pair (an s-d pair) and a routing scheme x, the degree of the optimality, represented by r_x, is defined as*

Table 1: Comparative results among *I-*, *K-*, *G- schemes* by the analytical evaluation.

Routing scheme	Optimality of path length
I–scheme	$2^{j-2}(3/2 + D/2) - 1$
K–scheme	$2^{j-2}(2 + D/2) - 1$
G–scheme	$(1 + \ell + D)^{j-2}(1 + \ell + 3D/2) - 1$

j: The level of the lowest level cluster which contains both source and destination nodes $(2 \leq j \leq h)$.

D: The maximum diameter among diameters of 1-level clusters.

ℓ: The maximum weight among weights of edges that join two different 1-level clusters.

a ratio of [the path length obtained by the scheme x] to [the distance between s and d on G]. Then, for given n (n $\geq$ 2) s-d pairs and a routing scheme x, the degree of the optimality is defined as an average value of n r_x's.

4.1 Assumptions for Simulation

The simulation model is a hierarchical network G^3. To simplify the model, the following H.1 and H.2 are assumed.

H.1 : All s-d pairs on the hierarchical network are generated at random.

H.2 : Each communication delay for each edge doesn't change during the simulation.

4.2 Experimental Data

The simulation consists of the following six steps.

S.1 : Create a hierarchical network G^3.

S.2 : Generate s-d pairs on the G^3.

S.3 : Compute the shortest paths, called the reference paths, for all s-d pairs on the G^3.

S.4 : Generate routing tables of the scheme for the G^3.

S.5 : Determine paths for the pairs by using routing tables.

S.6 : Compare the paths obtained at S.5 with the reference paths.

We explain the details of each step in order. At step S.1, a network graph $G=(V, E)$ is created at random. The number of nodes, represented by $|V|$, is 200, 400, or 600. The number of edges, denoted by $|E|$, is $1.4|V|$ or $1.6|V|$. Then, there are 6 different networks. Each edge in G is labeled by an integer among 1 and 5 at random.

Then, a hierarchical network G^3 is constructed on G by applying a clustering algorithm [2]. An initial clustering algorithm [2] is superior to conventional algorithms with respect to time complexity. The initial clustering algorithm constructs a clustered network G^2, which consists of $\sqrt{|V|}$ clusters, and each cluster has the $\sqrt{|V|}$ nodes on an average. The same algorithm can be used recursively to construct G^3 from G^2.

This experimental data is considered reasonable from a practical point of view. For example, MILNET, which is a typical computer network, included 250 nodes by 1987 and 700 nodes by 1991 respectively, and the total number of the edges is $1.45|V|$ by 1986 [9]. Three levels have been also pointed out as a typical number of levels in a hierarchical network [3, 9].

At step S.2, a set of s-d pairs is generated on G^3. The number of s-d pairs is set to 5% of the square of $|V|$. The determination of this number is based on the examination how many pairs are necessary to stabilize the value of r_x's. The examination was performed for 10 different hierarchical networks G^3's with $|V|=200$ and $|E|=1.6|V|$ by changing the number from 1% to 100%. As the result, 5% is sufficient to evaluate r_x's for each scheme.

At step S.3, the shortest path length for each s-d pair is calculated by Dijkstra's algorithm. At step S.4, routing tables at each node for a given scheme are constructed for G^3. At step S.5, a path for each s-d pair is determined by using a given scheme. Finally, at step S.6 for each s-d

Table 2: Simulation results based on the optimality of path length.

(a) $|E| = 1.4|V|$.

No. of Nodes	Ave(r_I)	Ave(r_K)	Ave(r_G)
200	1.15	1.31	1.71
400	1.17	1.36	1.80
600	1.22	1.43	1.95

(b) $|E| = 1.6|V|$.

No. of Nodes	Ave(r_I)	Ave(r_K)	Ave(r_G)
200	1.14	1.38	1.82
400	1.15	1.40	1.95
600	1.18	1.46	2.02

pair, the length of the path obtained at step S.5 and that of the shortest path are compared.

4.3 Optimality of Path Length

Table 2 (a) and (b) show the optimality of path length for the cases $|V|$=200, 400, 600, and $|E|$=1.4$|V|$ and $|E|$=1.6$|V|$, respectively. The first and second rows in Table 2 (a) and (b), which correspond to 200 and 400 nodes respectively, represent the average value of 10 different experiments. The third row, which corresponds to 600 nodes, represents the average value of 5 different experiments.

In Table 2 (a) and (b), Ave(r_x) (x =I, K, *and* G) denotes the average value of r_x's for x-scheme. From Table 2, it is clear that *I-scheme* is superior to *K-scheme* and *G-scheme* with respect to the optimality of path length on the average. The average path length obtained by *I-scheme* is within about 20% longer than the optimal path length. From Table 2, it is also observed that the difference of average values between *I-scheme* and the others increases when the number of nodes and the number of edges become larger.

Table 3 shows the worst cases for three routing schemes, in which r_x's are evaluated with respect to the optimal path length. In Table 3, Max(r_x) ($x = I$, K, G) denotes the maximum value of r_x's obtained by

Table 3: Simulation results of the worst case.

(a) $|E| = 1.4|V|$.

No. of Nodes	Max(r_I)	Max(r_K)	Max(r_G)
200	6.25	6.25	30.00
400	7.00	10.33	22.67
600	7.25	11.00	29.00

(b) $|E| = 1.6|V|$.

No. of Nodes	Max(r_I)	Max(r_K)	Max(r_G)
200	7.33	11.00	15.50
400	7.00	11.50	20.50
600	13.00	13.00	30.50

the simulation experiments.

From Table 3, *I-scheme* is superior to *K-* and *G-schemes* with respect to the worst cases of simulation experiments.

4.4   Routing Table Size

For the routing table size, the schemes have the same complexity $O(\log n)$ where n is the number of nodes in a network (see Table 4). Figure 7 shows a typical simulation result with respect to the size of routing table at each node. In the simulation experiments, the size of routing table at each node in *I-scheme* is about twice or three times the size of that in other schemes, that could be consider as a defect of *I-scheme*. However, as the mention before the size is comparable to the others. Moreover, the size of routing table at each node in *I-scheme* is by far smaller than that in a conventional (flat) routing scheme. It is observed that there is a tradeoff between the optimality of path length and the size of routing table.

5   Conclusion

In this paper, the time-space tradeoffs in hierarchical routing schemes are presented quantitatively. The time is viewed, here as message delay

Table 4: Analytical evaluation with respect to routing table size.

Routing scheme	Routing table size
$K - scheme$	$3C \log n$
$G - scheme$	$(3 + 4(\log n - 1))C$
$I - scheme$	$(3 + 4B(\log n - 1))C$

n: The number of nodes in a network.

B: The number of k-level border nodes in a k-level cluster $(1 \leq k < h)$.

C: The number of nodes in a k-level cluster.

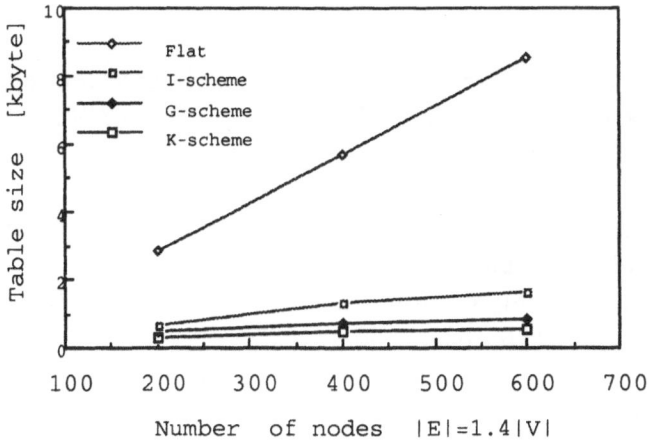

Figure 7: Simulation result with respect to routing table size.

between a source node and a destination node. With respect to the time, it is shown that *I-scheme* proposed by the authors is superior to other two schemes, *K-scheme* and *G-scheme*. As the space, *I-scheme* suffers from a deficiency of a few increases of routing table size compared with the other schemes. However, concerning the order of size of routing table, the schemes have the same complexity $O(\log n)$ where n is the number of nodes in a network. The simulation experiments show that the table size in *I-scheme* is comparable to it in the other schemes. The results represent the fundamental tradeoff between routing efficiency and routing table size.

Future researches include development of hierarchical routing that takes presence of faults and timely execution into consideration.

Acknowledgements

The author gratefully acknowledges useful suggestion from Prof. Miroslaw Malek of University of Texas at Austin. The comments of the referees are also appreciated.

References

[1] J. J. G. L. Aceves and N. Shacham, "Analysis of routing strategies for packet radio networks," *Proceedings of the IEEE INFOCOM'85*, pp. 292–302, 1985.

[2] K. Ishida, J. Miyao, T. Kikuno, and N. Yoshida, "On analytical evaluation of hierarchical routing schemes for large computer networks –On the optimality of path length–," *Trans. IEICE Japan*, vol. J71-A, no. 8, pp. 1576–1584, 1988.

[3] L. Kleinrock and F. Kamoun, "Hierarchical routing for large networks: Performance evaluation and optimization," *Computer Networks*, vol. 1, pp. 155–174, 1977.

[4] C. V. Ramamoorthy and W. T. Tsai, "An adaptive hierarchical routing algorithm," *Proceedings of the COMPSAC 83*, pp. 93–104, 1983.

[5] M. Malek, "Responsive systems: The challenge for the nineties," *Proceedings of the EUROMICRO'90, 16th Symp. on Microprocessing and Microprogramming, Keynote Address*, Microprocessing and Microprogramming 30, pp. 9–16, North-Holland, 1990.

[6] K. Ishida, *A study on hierarchical routing scheme based on network clustering for large computer networks*. PhD thesis, Faculty of the Systems Engineering, Hiroshima University, 1989.

[7] T. Kikuno, J. Miyao, K. Ishida, and N. Yoshida, "A cluster-based approach to fault-tolerant computer networks," *Proceedings of the ISCAS 86*, pp. 630–633, 1986.

[8] G. S. Lauer, "Hierarchical routing design for SURAN," *Proceedings of the ICC'86*, pp. 4.2.1–4.2.10, 1986.

[9] J. Seeger and A. Khanna, "Reducing routing overhead in a growing DDN," *Proceedings of the IEEE MILCOM'86*, pp. 15.3.1–15.3.13, 1986.

Work in Progress

Fault-Tolerance Support for Responsive Computer Systems

R. D. SCHLICHTING

Department of Computer Science

The University of Arizona

Tucson, Arizona 85721, USA

Abstract

A number of projects related to supporting fault-tolerance in responsive systems are described. One area of work concerns the development of appropriate system abstractions. The focus here has been on designing and implementing Consul, a communication substrate for constructing fault-tolerant distributed programs. The other area concerns improving programming language support. The main projects in this area include a high-level distributed programming language called FT-SR, and a version of the Linda coordination language with fault-tolerance enhancements.

Key Words: Dependability, system abstractions, programming languages.

1 Introduction

Our research is concerned with increasing computer system *dependability* that is, the basic trustworthiness of a computer system that allows people to rely on the service it delivers [1]. Within this general area, we have been concentrating on developing techniques and support systems for constructing *fault-tolerant, distributed programs*; such a program is a distributed program that can continue executing despite processor or network failures. These programs are often used for responsive systems that have real-time requirements as well; examples of this type of system include air traffic control, nuclear reactors, and aircraft flight systems.

Over the past few years, we have been conducting investigations in a number of different areas related to fault-tolerant, distributed programming. Two in particular have seen substantial attention. One is the development of appropriate system abstractions, especially those concerned with group-oriented interprocess communication; the other is investigating enhanced programming language support for this type of programming. Both are relevant for responsive systems, although our work has not explicitly considered real-time requirements up to now. We describe each of these areas in turn, with an emphasis on continuing work.

2   System Abstractions

Developing appropriate system abstractions has been the basis for many of the advances in software technology for highly dependable systems over the years. For example, transactions were developed to ease the task of constructing database applications, while stable storage provides similar benefits for a wide variety of fault-tolerant applications. Hence, a primary challenge in the area of fault-tolerant programming is to develop abstractions that are powerful enough to be useful, yet simple enough that they can be implemented efficiently at an appropriate level of the system hierarchy.

Our particular focus has been on developing *fault-tolerant services* for the *replicated state machine* approach to constructing fault-tolerant software. This approach is based on taking a server implemented as a state machine and replicating it to mask failures [2]. Examples of useful fault-tolerant services for this approach include multicast services to deliver messages to a collection of processes reliably and in some consistent order, membership services to maintain a consistent system-wide view of which processes are functioning and which have failed, and recovery services to recover a failed process. A large amount of research has been performed in areas related to this approach, including development of new algorithms and systems (e.g., [3, 4, 5, 6].)

Over the past two years, we have designed and implemented Consul, a unified collection of protocols that provide these services. This collection of protocols, which forms a communication substrate upon which fault-tolerant applications can be built, provides support to manage redundancy, failures, and recovery in a distributed system. Specifically, the fault-tolerance support includes process failure detection, restart of failed processes, and reliable communication between processes. Support for general distributed processing is included in this substrate as well. This support includes interprocess communication within a group of processes and different kinds of consistent orderings. The thrust of the work has been both to devise new algorithms for the various protocols, as well as to explore new system structuring techniques.

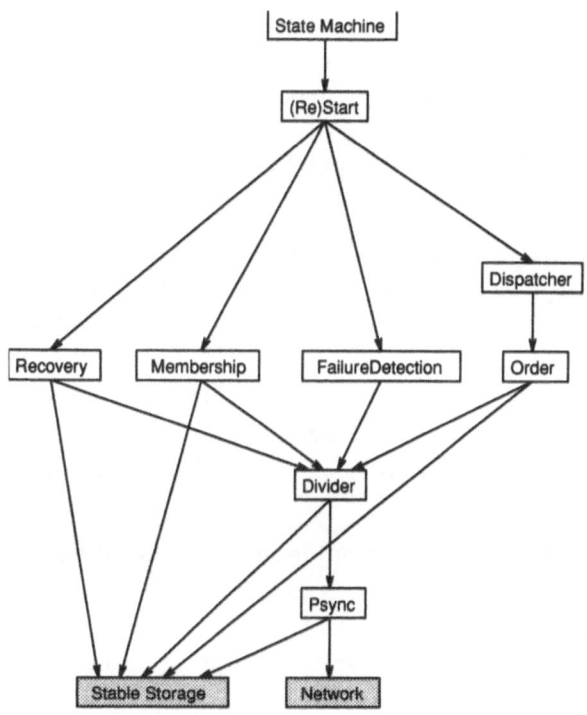

Figure 1: Structure of Consul

The protocol structure of Consul is illustrated in Figure 1; in this figure, there is an arrow from protocol u to protocol v if u uses v to implement its functionality. At the base of the substrate is the stable store service, which is used by all other services to periodically checkpoint their states. Psync is the main communication mechanism of the substrate; it provides a group-oriented interprocess communication mechanism in the form of a multicast facility that maintains the partial (or causal) order among messages. The order protocol is chosen from a suite of different and independent protocols, each providing a different kind of consistent message ordering on top of Psync; the most novel of these is a "semantic-dependent" order that exploits the partial order provided by Psync to allow more application concurrency. The failure detection protocol monitors processes for potential failures, while the membership protocol maintains a consistent system-wide view of group membership. The membership protocol again exploits partial order, resulting in flexible and efficient handling of simultaneous failures and recoveries. The recovery protocol deals with restoring the state of the recovering process to the current state and cooperates with membership to incorporate the process into the group; recovery makes use of the checkpoints stored by the different protocols as well as the message retransmission mechanism of Psync. Finally, (re)start, dispatcher, and divider are used primarily for configuration purposes.

An initial version of Consul has been fully implemented. The code consists of approximately 10,000 lines of C, of which 3,500 is Psync. The implementation vehicle for the substrate is the x-kernel, an operating system kernel designed explicitly for experimenting with communication protocols [7]. Initial experiments with two applications, a replicated directory object and a distributed word game, have shown that exploitation of the partial ordering can, in fact, result in a performance advantage.

Several papers have been written describing Consul and its algorithms. In [8], the membership protocol is described and its correctness argued. The entire system is the subject of [9] and a recently completed Ph.D. dissertation [10]. Our experiences attempting to modularize this type of protocol is the subject of another paper [11].

Work continues on Consul and other aspects of fault-tolerant protocols. For example, we are currently attempting to decompose Psync into multiple smaller protocols in order to better isolate and understand its various properties. More recently, we have also begun exploring a fundamentally new model for protocols of this type in which protocols do not function as message filters, but rather operate concurrently on a shared pool of messages; this "data centered" orientation is based on the observation that many of these protocols are involved with transforming the relationship between messages, something difficult to do with a filter. Although just underway, initial results seem to indicate that it may be possible to decompose and modularize fault-tolerant protocols using this model to a greater degree than previously thought.

3   Programming Language Support

Over the past ten years, numerous high-level programming languages and notations have been developed specifically for writing programs to execute on distributed architectures. These include languages such as Ada [12], Modula-2 [13], and SR [14]. However, while incorporating many interesting ideas about such things as synchronization, distribution, and encapsulation, little attention has been paid in these languages to supporting the needs of fault-tolerant programming. This lack of appropriate language support is especially true for systems-level programming, since languages that do explicitly focus on fault-tolerant programs are typically oriented towards constructing applications.

We have been pursuing two efforts aimed at improving this situation. One is the development of FT-SR, a programming language based on SR that is oriented towards writing fault-tolerant, distributed systems. In addition to its focus on systems-level programming, this language is also unique in that it is specifically designed to support equally well *any* of the multiple programming paradigms that have been developed for this type of system. In addition to the replicated state machine approach mentioned above, these paradigms include the use of *restartable actions* [15] and the *object-action model* [16]. The development of a language supporting multiple paradigms for fault-tolerant programming is, in our view, analogous to the evolution of standard distributed programming languages from those that supported only a single type of synchronization to those such as SR that support multiple approaches. The orientation towards supporting multiple paradigms distinguishes FT-SR from other languages [17, 18, 19], language extensions [20, 21], and language libraries [3, 22, 23] related to fault-tolerance, which are typically oriented around a particular paradigm.

Over the past year, we have prepared an initial design for the language. In doing this, we first developed a new model to serve as a common basis for all of these disparate paradigms. The fundamental abstraction in this model is a *fail-stop atomic object*; such objects behave as atomic objects unless their implementation assumptions are violated, in which case notification is provided. Fail-stop atomic objects are combined with standard fault-tolerance techniques like replication and recovery to provide a simple model for structuring and implementing fault-tolerant, distributed systems.

FT-SR is designed around the concept of fail-stop atomic objects. However, given its role as a systems programming language, we do not provide these objects directly in the language, but rather include features that allow them to be easily implemented using any of the existing techniques. To this end, the language has provisions for encapsulation based on SR resources, resource replication, recovery protocols, synchronous failure notification when performing interprocess communication, and a mechanism for asynchronous failure notification. A paper outlining the initial design has been completed [24]. Work is progressing on implementing FT-SR using as a starting point an existing version of SR that has been ported to the x-kernel. The resulting language will execute standalone on a network of Sun workstations.

The second language-oriented effort is investigating the use of more non-traditional languages and notations for performing fault-tolerant, distributed programming. The basic goal of this work is to determine whether some of these languages support programming models or features that facilitate the task of writing this type of program.

The bulk of our work so far has concentrated on Linda, a system for constructing distributed applications based on a communication abstraction known as tuple space [25]. In particular, we have developed a simple technique for enhancing the fault-tolerance of programs that use the "bag-of-tasks" programming paradigm, a class of programs for which Linda is especially well-suited. This paradigm, in which the problem space is divided up and parceled out to processes as independent subtasks, is a common technique for constructing parallel solutions to a wide variety of problems, such as matrix multiplication, DNA sequencing, and the traveling salesman problem.

Our technique, which is outlined in [26], is based on extending Linda with a *conditional atomic tuple swap* operation. This operation facilitates construction of a *fault-tolerant bag-of-tasks* in which tasks being executed on processors that crash can easily be reallocated to a new process on a functioning processor. In addition to being minor, our extensions also have the virtue of being easy to build using previously-developed algorithms for implementing a stable tuple space, i.e., a tuple space that can survive processor failures [27].

We plan to continue our research involving Linda along several lines. One will involve implementing the extensions outlined above; in addition to the swap operation, this effort will also include implementing a stable tuple space since no realization of the design alluded to above actually exists. Another will involve expanding into more general investigations of fault-tolerance and Linda. For example, we intend to investigate exploiting Linda's unique properties to realize some of the fault-tolerant programming paradigms mentioned above. This will most likely lead us to propose further Linda extensions, which will be implemented and tested in turn. To lend concreteness, we plan on performing this work in the context of realistic applications that require fault-tolerance. For example, we are currently beginning a collaborative effort with researchers at GTE Laboratories in Massachusetts that will investigate use of these Linda techniques for implementing enhanced telephone services such as call forwarding and call waiting.

In a related project involving collaboration with Japanese colleagues, we are also adapting AG, a programming language developed at Tokyo Institute of Technology [28], in much the same way by extending it to support recovery and replication. Although other languages have been developed with these techniques in mind, AG appears to offer unique advantages that we hope to exploit to develop a better approach. These advantages derive from the fact that AG is based on an *attribute grammar* formalism, and as such, shares its advantages of a declarative style, separation of semantic and syntactic definitions, and a functional foundation. To our knowledge, no one has yet attempted to exploit the special characteristics of attribute grammars—or even functional programming languages—in the construction of fault-tolerant software.

4 Conclusions

An important aspect of responsive computer systems in ensuring resilience to failures. Here, we have briefly described several efforts aimed at making this difficult task easier, including a system that provides enhanced communication facilities and work on programming language support for fault-tolerant programming. Future plans in these areas have also been outlined.

Acknowledgments

This work supported in part by NSF Grant CCR-9003161 and ONR Grant N00014-91J-1015.

References

[1] J. C. Laprie, ed., *Dependability: Basic Concepts and Terminology.* Vienna: Springer-Verlag, 1992.

[2] F. Schneider, "Implementing fault-tolerant services using the state machine approach: A tutorial," *ACM Computing Surveys*, vol. 22, pp. 299–319, Dec 1990.

[3] K. Birman, A. Schiper, and P. Stephenson, "Lightweight causal and atomic group multicast," *ACM Transactions on Computer Systems*, vol. 9, pp. 272–314, Aug 1991.

[4] F. Cristian, "Agreeing on who is present and who is absent in a synchronous distributed system," *Proc. of 18th International Conference on Fault-tolerant Computing*, (Tokyo), pp. 206–211, Jun 1988.

[5] D. Powell, D. Seaton, G. Bonn, P. Verissimo, and F. Waeselynk, "The Delta-4 approach to dependability in open distributed computing systems," *Proc. of 18th Symposium on Fault-Tolerant Computing*, (Tokyo), Jun 1988.

[6] P. Verissimo, L. Rodrigues, and M. Baptista, "Amp: A highly parallel atomic multicast protocol," *SIGCOMM'89*, (Austin, TX), pp. 83–93, Sep 1989.

[7] N. C. Hutchinson and L. L. Peterson, "The x-kernel: An architecture for implementing network protocols," *IEEE Transactions on Software Engineering*, vol. 17, pp. 64–76, Jan 1991.

[8] S. Mishra, L. L. Peterson, and R. D. Schlichting, "A membership protocol based on partial order," *Dependable Computing for Critical Applications 2* (J. Meyer and R. Schlichting, eds.), pp. 309–331, Vienna: Springer-Verlag, 1992.

[9] S. Mishra, L. L. Peterson, and R. D. Schlichting, "Consul: A communication substrate for fault-tolerant distributed programs," Tech. Rep. TR 91-32, Dept of Computer Science, University of Arizona, Tucson, AZ, 1991.

[10] S. Mishra, *Consul: A Communication Substrate for Fault-tolerant Distributed Programs.* PhD thesis, Dept of Computer Science, University of Arizona, Tucson, AZ, 1991.

[11] S. Mishra, L. L. Peterson, and R. D. Schlichting, "Experience with modularity in Consul," Tech. Rep. 92-25, Dept of Computer Science, University of Arizona, Tucson, AZ, 1992.

[12] U. S. Department of Defense, *Reference Manual for the Ada Programming Language.* Washington D.C., 1983.

[13] N. Wirth, *Programming in Modula-2.* New York: Springer-Verlag, 1982.

[14] G. R. Andrews, R. A. Olsson, M. A. Coffin, I. Elshoff, K. Nilsen, T. Purdin, and G. Townsend, "An overview of the SR language and implementation," *ACM Transactions on Programming Languages and Systems*, vol. 10, pp. 51–86, Jan. 1988.

[15] B. Lampson, "Atomic transactions," *Distributed Systems— Architecture and Implementation*, pp. 246–265, Berlin: Springer-Verlag, 1981.

[16] J. Gray, "An approach to decentralized computer systems," *IEEE Transactions on Software Engineering*, vol. SE-12, pp. 684–692, Jun 1986.

[17] C. Ellis, J. Feldman, and J. Heliotis, "Language constructs and support systems for distributed computing," *ACM Symposium on Principles of Distributed Computing*, pp. 1–9, August 1982.

[18] R. J. LeBlanc and C. T. Wilkes, "Systems programming with objects and actions," *Proc. of 5th International Conference on Distributed Computing Systems*, (Denver, Colorado), pp. 132–139, May 1985.

[19] B. Liskov, "The Argus language and system," *Distributed Systems: Methods and Tools for Specification, Lecture Notes in Computer Science, Volume 190* (M. Paul and H. Siegert, eds.), ch. 7, pp. 343–430, Berlin: Springer-Verlag, 1985.

[20] R. Cmelik, N. Gehani, and W. D. Roome, "Fault Tolerant Concurrent C: A tool for writing fault tolerant distributed programs," *Proc. of 18th International Symposium on Fault-Tolerant Computing*, (Tokyo), pp. 55–61, June 1988.

[21] J. Knight and J. Urquhart, "On the implementation and use of Ada on fault-tolerant distributed systems," *IEEE Transactions on Software Engineering*, vol. SE-13, pp. 553–563, May 1987.

[22] F. Panzieri and S. K. Shrivastava, "Rajdoot: A remote procedure call mechanism supporting orphan detection and killing," *IEEE Transactions on Software Engineering*, vol. SE-14, pp. 30–37, Jan 1988.

[23] M. Herlihy and J. Wing, "Avalon: Language support for reliable distributed systems," *Proc. of 17th International Symposium on Fault-Tolerant Computing*, (Pittsburgh, PA), pp. 89–94, July 1987.

[24] R. D. Schlichting and V. T. Thomas, "A multi-paradigm language for programming fault-tolerant distributed systems," *Proc. of 11th Symposium on Reliable Distributed Systems*, (Houston, TX), pp. 222–229, Oct 1992.

[25] S. Ahuja, N. Carriero, and D. Gelernter, "Linda and friends," *IEEE Computer*, vol. 19, pp. 26–34, August 1986.

[26] D. Bakken and R. D. Schlichting, "Tolerating failures in the bag-of-tasks programming paradigm," *Proc. of 21st Symposium on Fault Tolerant Computing*, (Montreal, Canada), pp. 248–255, Jun 1991.

[27] A. Xu and B. Liskov, "A fault-tolerant distributed implementation of Linda," *Proc. of 19th Symposium on Fault-Tolerant Computing*, (Chicago, IL), Jun 1989.

[28] Y. Shinoda and T. Katayama, "Attribute grammar based programming and its environment," *Proc. of 21st Hawaii International Conference on System Sciences*, (Kailu-Kona), pp. 612–620, Jan 1988.

Position Paper: Responsive Airborne Radar Systems

L. SHA, J. LEHOCZKY, M. BODSON

SEI/CMU

P. KRUPP, C. NOWACKI

The MITRE Corporation

Abstract

This paper is an initial work-in-progress report on a joint effort between MITRE and the SEI in the development of an experimental prototype of a responsive airborne radar system. This effort employs: (1) analytic redundancy to guard against application level software errors, (2) runtime isolation and fault containment techniques to guard against programming system level software errors such as illegal addressing, and (3) generalized rate monotonic scheduling techniques to guard against timing errors.

Key Words: Real-time, fault-tolerance, analytical redundancy, tracking algorithms, rate-monotonic scheduling, hard real-time, software errors.

1 Introduction

Airborne surveillance radar systems define an important class of real-time systems. The processing that is commonly required of a typical surveillance radar is divided into pulse functions, detection functions, and tracking functions.

The pulse functions are generally highly regular, deterministic functions whose output size is pre-determined. Hard real-time deadlines are imposed on these tasks based on the specific pulse repetition rate and slant time of the radar. The detection functions operate on data from many pulses to identify returns from targets (detections). The functions performed are still highly deterministic, and the real-time constraints are

linearly related to the deadlines in the pulse processing. The size of
the output of the detection processing, however, is not deterministic,
because it is impossible to predict how many targets will be detected.
This information is then processed to track targets, using filtering functions that will predict future target locations. The track processing also
must meet real-time constraints, but the difficulty is that the amount of
processing to be done will necessarily vary depending on the particular
target scenario. As the workload varies with the target environment,
the system must be capable of operating in a gracefully degraded mode
when the workload becomes too great to meet the deadlines.

The processing requirements of these systems are very high, often in
the range of billions of operations per second. These systems usually
must be implemented on a distributed/parallel computer architecture
with highly complex software, frequently upgraded as the mission requirements change. As with any complex software system, the formal
verification of such radar tracking software systems is currently not practical. Nevertheless, such systems must execute predictably and reliably
despite overloads or the activation of latent software errors. The importance of overcoming overloads and latent software errors is that they,
rather than hardware faults, dominate the mean time to failure in the
operation of a highly complex airborne surveillance system. In the rest
of the paper, we shall assume that the hardware fault tolerance problem
is handled by other means.

2   A Fault Model

From our initial work, we found that it was useful to classify software
errors into three types for the purpose of detection and recovery.

1. Inaccuracy: In the context of our applications, these are tracking
   errors, which are a function of the quality of the data and the so-
   phistication of the tracking algorithm. Due to the nature of the
   application, such errors can only be reduced but not eliminated.
   Design or implementation errors in software development can also
   contribute to tracking errors.

2. Timing faults: These typically occur in the form of missed deadlines. While software design and implementation errors may lead

to timing faults, a major source of timing faults is the complexity of the algorithms. The difficulty in tracking applications is that sophisticated algorithms may reduce the number of tracking errors but contribute to timing faults.

3. Programming system faults: These are those serious software faults that may crash the system, for example, illegal addresses or data, exhausting available buffers, monopolizing the I/O channels and/or CPU.

Both simple and complex algorithms will have tracking errors, but the simple algorithms generally will have more tracking errors. The complex software uses sophisticated algorithms which, in principle, should have comparatively fewer tracking errors. However, they are prone to timing faults, since their worst case timing behavior is exponential. One may arbitrarily cut short the computation time of such algorithms. However, one can no longer say that the quality the estimation of the complex software is better than that of the simple software, should one arbitrarily cut short its computation time. Finally due to the lack of experience with such systems, the software implementing complex algorithms is likely to have more bugs than the simple algorithmic system which has been tested over a long time. These errors can lead to programming system faults. Figure 1 is a model which compares the characteristics of a simple software system with those of a complex software system.

The software architecture used to deal with these faults is known as the Simplex Software Architecture[1]. This architecture employs: (1) analytic redundancy to guard against application level software errors, (2) runtime isolation and fault containment techniques to guard against programming system level software errors such as illegal addressing, and (3) generalized rate monotonic scheduling techniques to guard against timing errors. In this paper, we will focus on the use of analytic redundancy to contain application level software errors. Figure 2 is the conceptual model that illustrates the combined use of scheduling, runtime fault containment and analytic redundancy to improve the overall functional performance and reliability.

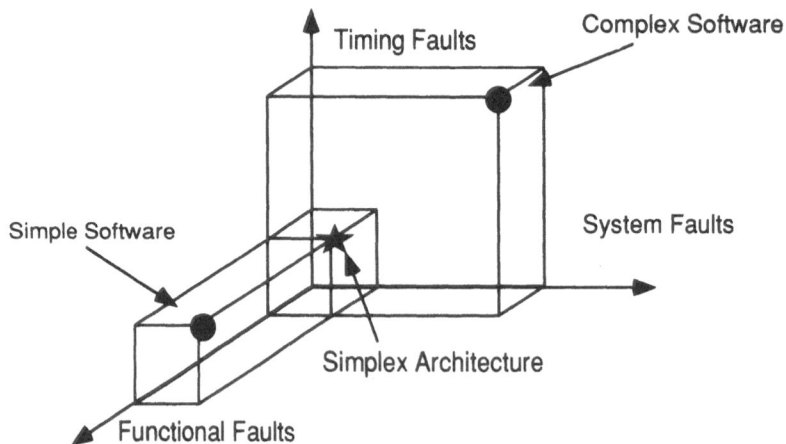

Figure 1: A Fault Model

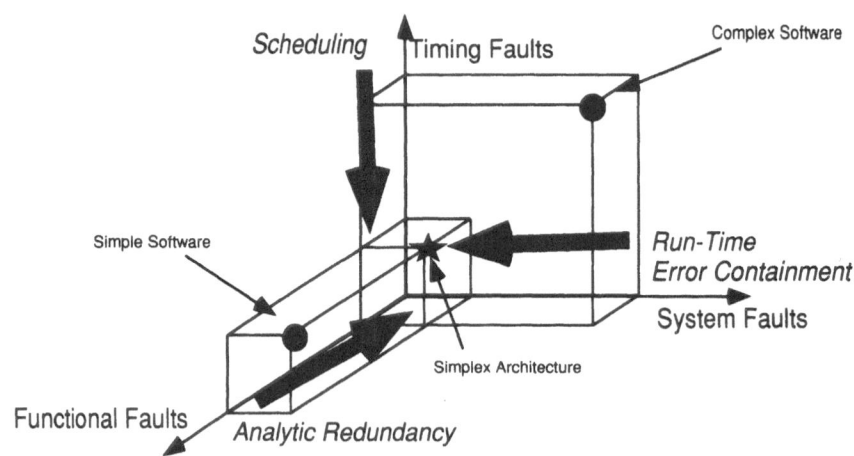

Figure 2: Fault Management

3   The Concept of Analytic Redundancy

To deal with software faults, some form of redundancy in computation is needed. Direct redundancy uses different programs to compute the same results so that voting or mid-value selection can be used. An example of direct redundancy for software fault tolerance is N-version programming[2]. The problem with N-version programming is that it is very expensive to implement and there is no assurance that errors resulting from independent developments will be independent[3]. Indeed, programmers with similar backgrounds can make common mistakes, even in separate development efforts. When analytic redundancy is used, different computations can produce different results that are functionally related. The recoverable block approach[4] uses simpler alternatives as a back-up to the primary one. If the primary cannot pass the acceptance test, the simpler alternative will run and try to pass the test. The simpler alternative is a form of analytic redundancy. A problem with recovery blocks often found in practice is the difficulty of writing tight acceptance tests. It is difficult to detect a legal but incorrect answer without knowing the characteristics of the correct answer.

To overcome this problem, our approach is to develop a trusted simple software system which will give us a baseline answer plus a set of confidence assertions. A confidence assertion is a generalization of the statistical concept of confidence interval, and we will illustrate this concept in the context of tracking. The complex software is not trusted. Its outputs must be consistent with the confidence assertions produced by the simple software or they will be discarded. Furthermore, we are not even able to trust with the computation process employed by the complex software. This process may take too long for real-time applications or even crash the entire application. We will return to this subject in the next section when we present our fault model.

There is a large family of tracking algorithms and filtering algorithms with distinctively different properties along the dimensions of logical complexity, timing complexity and accuracy. "Greedy" algorithms perform local optimizations each step along the way. They tend to be low in logical complexity and timing complexity. They are simple to implement and could be verified formally in many instances. Their worst case time complexity is typically a low order polynomial, e.g., $O(n^2)$, where n represents the maximum of the number of returns per sweep and the number of tracks in the track file. Unfortunately, they also tend to generate larger estimation errors compared with more sophisticated algorithms. The larger errors may lead to problems such as "ghost targets." Sophisticated algorithms can provide smaller estimation errors and can have fairly good average execution time. Unfortunately, they are more complex logically and their logical complexity tends to grow over time due to the addition of optimizations and heuristics for various important special cases. Furthermore, despite their good average timing behavior, their worst case timing complexity is typically exponential in nature. Indeed, the dichotomy between average and worst case timing behavior in combinatorical problems is the rule rather than exception. For example, the well known simplex algorithm for linear programming exhibits such behavior.

The basic idea of the simplex architecture[1] for airborne radar systems is to combine the strengths of the simple and the complex algorithms. That is, we will combine algorithms with low logical complexity and timing complexity but higher estimation errors together with algorithms with higher logical complexity and timing complexity but lower estimation errors. The simple algorithms will provide a defense against timing failures due to the triggering of worst case execution timing behavior and logical failures due to the activation of latent design and implementation errors. The complex algorithm will provide the desired accuracy when it is active. In the following discussion, the software that implements the complex (simple) algorithms will be referred to as the complex (simple) software.

4 Experimental Set Up

In the current experiment, we use a uni-processor. We plan to parallelize the tracking computation in the future. Under our simplex software architecture, we create three processes, the management process which spawns two child processes, the simple software process and the complex software process. The scope of the task scheduling is global. That is, the task with the highest priority in any process will run. All three

---

[1]The simplex algorithm for linear programming and the simplex architecture for software fault tolerance have no relationship.

processes have multiple tasks scheduled by the generalized rate mono-tonic algorithm. However, the scheduling issue is not the focus of this paper.

Currently, the simple software is based on the nearest neighbor (NN) algorithm, and the complex software is the joint probability data association algorithm (JPDA)[5]. We plan to replace it with a more sophisticated algorithm such as the multiple hypothesis tracking algorithm[5]. The JPDA has a complexity that is higher than NN by a factor of 2^T, where T is the number of tracks within the gate. We plan to use the generalized rate monotonic scheduling (GRMS) approach as the baseline for dealing with timing faults. GRMS includes the original rate monotonic scheduling algorithm[6] and the subsequent generalizations to the scheduling theory reviewed in[7, 8]. We will further extend the approach whenever the need arises. Furthermore, we will use the simplex software architecture[1] to employ analytic redundancy to defend against the inevitable transient overloads and latent software errors.

The simple software will always run to completion and pass all its results to the management process. Based on the known error model of the simple software, the complex software will be invoked to reduce estimation errors when the simple software encounters a complex situation that it is known to be difficult to handle. To guard against semantic faults, the management process uses two sets of checks. One is the customary sanity check based on domain knowledge. The other is a set of confidence assertions provided by the simple software. Confidence assertions are a generalization of the statistical concept of the confidence interval. For example, a confidence assertion can be in the form of "there are two tracks A and B crossing at position x, y, z within a radius of r feet. The two emerging tracks are either A or B (i.e. one is unable to distinguish which is which after the crossing.)" In other words, confidence assertions are what the simple software can be sure of with a high degree of confidence. The answer produced by the complex software must be produced in time and be consistent with both the sanity checks and with the confidence assertions of the simple software. Otherwise, the complex software will be reset and restarted in a new state. The answer produced by simple software will be used instead.

To guard against resource, timing and programming system faults, the complex software process has a fixed number of buffers, no I/O privileges and communicates with others only by messages, nor can it alter the task priorities. The management process will handle the system error signal should the complex software fail. To guard against timing faults of the complex software, the management process runs a high priority task that will wake up as the deadline of any task in the complex software arrives. Alternatively, we can build these timing checks into the runtime scheduler. There are several interesting questions we seek to answer in our planned experiment.

- Is the complex software being invoked too frequently or too infrequently for the purpose of reducing estimation errors?

- What is the probability that the complex software's answer is correct but is rejected?

- What is the probability that the complex software's answer is wrong but is accepted?

- What is the performance improvement that can be obtained by using complex software with residual errors?

We are currently investigating experimentally the tracking and timing performance of trackers based on both the NN and JPDA algorithms. This is being done by defining tracking scenarios and then evaluating the performance of the algorithms through a tracking system simulation. The preliminary results indicate that JPDA indeed performs better in a majority of cases but at a much higher computational cost. These tenative results also indicate that simple confidence assertions can be implemented that in real-time reliably determine when a fall back to NN is appropriate for a significant class of JPDA failures. However, the sample size is currently too small for us to make firm conclusions about these results.

5   Conclusion

The processing requirements of modern airborne radar systems are very high, often in the range of billions of operations per second. These systems must usually be implemented by a distributed/parallel computer architecture with highly complex software, frequently upgraded as the mission requirements change. As with any complex software system, the formal verification of such software systems is currently not practical. Nevertheless, such systems must execute predictably and reliably despite overloads and even the activation of latent software errors.

This paper describes an initial approach to the application of generalized rate-monotonic scheduling techniques coupled with analytic redundancy to the design of such a prototype radar system. The basic idea is to combine the strength of simple and complex algorithms. The simple algorithms will provide a defense against timing failures triggered by the worst case execution timing behavior and logical failures due to the activation of latent design and implementation errors. The complex algorithm will provide the desired accuracy during normal operation.

Acknowledgement

This work is supported in part by a grant from the Office of Naval Research, N00014-92-J-1524 and in part by the SEI and MITRE corporation.

We want to thank Ragunathan Rajkumar of the SEI and Sean O'Neil, Ken King and Bill Eastman of MITRE Corporation for their suggestions, comments and ideas.

References

[1] L. Sha, J. P. Lehoczky, and M. Bodson, "The simplex architecture: Using analytical redundancy for software fault tolerance," *Proc. of The First International Workshop on Responsive Systems*, Oct. 1991.

[2] A. Avizienis, "The n-version approach to fault tolerant software," *IEEE Transaction on Software Engineering*, no. SE–11, 1985.

[3] J. C. Knight and P. E. Ammann, "Design fault tolerance," *Engineering and System Safety*, vol. 32, 1991.

[4] B. Randell, "System structure for software fault tolerance," *IEEE Transactions on Software Engineering*, vol. SE–1, 1975.

[5] S. Blackman, *Multiple-Target Tracking with Radar Applications*. Artrech House, 1986. ISBN 0-89006-179-3.

[6] C. L. Liu and L. J. W., "Scheduling algorithms for multiprogramming in a hard real time environment," *JACM*, vol. 20 (1), pp. 46 – 61, 1973.

[7] L. Sha and J. B. Goodenough, "Real-time scheduling theory and Ada," *IEEE Computer*, Apr. 1990.

[8] J. P. Lehoczky, L. Sha, J. K. Strosnider, and H. Tokuda, "Fixed priority scheduling theory for hard real-time systems," *Foundations of Real-Time Computing: Scheduling and Resource Management*, Kluwer Academic Publishers, 1991.

Overview of an Integrated Toolset Under Development for the CSR Paradigm *

I. LEE, S. DAVIDSON

Department of Computer and Information Science

University of Pennsylvania

Philadelphia, PA 19104

Abstract

The potential high cost associated with an incorrect operation of real-time systems has created a demand for a rigorous framework in which various design alternatives can be specified and analyzed before implementation. In addition, most real-time systems are costly to prototype, requiring careful prediction of timing properties before implementation and evaluation of design alternatives. In this paper, we briefly overview our approach to the specification and analysis of distributed real-time systems and a set of tools that are being developed. We then discuss various extensions that are being made to improve the applicability of the approach.

Key Words: Real-time systems, specification and analysis, real-time process algebra, communicating shared resources, toolset.

1 Introduction

The correctness of a real-time system depends not only on how concurrent processes interact, but also the time at which these interactions occur. This temporal correctness can be quite critical, as real-time systems are frequently used in such environments as manufacturing, robotics, avionics, communications, and medicine. In such applications, a system failure can often result in a high financial loss, or even the

---

*This research was supported in part by ONR N00014-89-J-1131 and NSF CCR90-14621.

loss of life. Thus, there is an increasing demand for the development of reliable, i.e., correct and safe, real-time systems.

Reliability in real-time systems can be improved through use of formal methods in the specification and analysis of real-time systems. Formal methods treat system components as mathematical objects and provide mathematical models to describe and predict the observable properties and behaviors of these objects. There are several tenets of using formal methods for the specification and analysis of real-time systems. They are as follows:

- early discovery of ambiguities, inconsistencies and incompleteness in informal requirements;

- automatic or machine-assisted analysis of the correctness of specifications with respect to requirements; and

- evaluation of design alternatives without expensive prototyping.

There has recently been a spate of progress in the development of real-time formal methods. Much of this work has fallen into the traditional categories of untimed systems; for example, temporal logics, assertional methods, net-based paradigms, process algebras, and programming languages. In this paper, we briefly overview our approach which is based on a real-time programming language and a real-time process algebra.

2   The CSR Paradigm

We have been developing a formal framework for reasoning about the temporal properties of real-time systems. Our framework includes a high-level real-time language, a configuration language, an abstract model of computation based on real-time process algebra, and a notion of equivalence of terms in the model.

Since the timing behavior of a real-time system depends not only on delays due to process synchronization, but also on the availability of shared resources, the computation model must include a notion of resources and how they can be shared as well as a notion of processes and synchronization. These notions are partially addressed in real-time models and scheduling theory, but not adequately combined. While most real-time models capture delays due to process synchronization, they abstract out resource-specific details by assuming idealistic operating environments. On the other hand, while scheduling theory captures the notion of resources, it ignores the effect of process synchronization except for simple precedence relations between processes. The salient aspect of our model is that it integrates these two notions.

To help bridge the gap between abstract computation models and implementation environments, we have developed a real-time language called *Communicating Shared Resources*, or CSR. CSR's underlying computational model is *resource-based*, where a resource may be a processor, an Ethernet link, or any other constituent device in a real-time system. At any point in time, each resource has the capacity to execute an action consisting of only a single event or particle. However, a resource may host a set of many processes, and at every instant, any number of these processes may compete for its availability. That is, on a single resource, the actions of multiple processes must be interleaved; "true" parallelism may take place only *between* resources. To arbitrate between competing events, CSR employs a priority-ordering.

CSR syntactically resembles variants of real-time CSP. However, it also has the capacity to specify many constructs commonly found in real-time systems, such as timeouts, deadlines, periodic processes, temporal scopes and exception-handling [4]. CSR also incorporates several of the features of a configuration language, in that processes must be explicitly assigned to the resources on which they reside.

CSR supports a natural, high-level description of real-time systems, and its semantics captures the temporal properties of prioritized resource interaction. However, the CSR language is far too complex to be treated as a process algebra, and thus does not easily lend itself to an equational characterization. To remedy this, we have developed the Calculus for Communicating Shared Resources, or CCSR [1, 2]. Strongly influenced by SCCS, CCSR is a priority-sensitive process algebra that uses a synchronous form of concurrency, and possesses a term equivalence based on strong bisimilarity. Thus CCSR provides the ability to perform equivalence proofs by syntactic manipulation. Also, since its prioritized, strong equivalence is a congruence, it allows us to reason about a term's behavior when it is embedded in a real-time "context." CSR and CCSR share the same basic computational model, in that they are both resource-based and rely on a priority arbitration scheme to resolve resource contention.

2.1 A Toolset Under Development

Based on our formal framework, we are developing a set of tools for
the specification and analysis of real-time systems. The specification is
written in CSR because it supports the high-level, natural description of
real-time systems. It then is translated into a real-time process algebra
called CCSR [3]. Based on CCSR, there are three tools that can be used
to analyze process behavior: a simulator, an analyzer and a mechanical
theorem prover.

The first tool, a hierarchical graphical interface and simulator, is based
on the graphical specification language being developed and the labeled
transition system of CCSR. The simulator executes the specification
with an environment which can either be specified by another process
or interactively provided by a human. The simulator is useful for testing
whether the specification is reasonable. Many obvious short-comings or
errors can easily be detected during the execution of the specification.

The second tool, an analyzer, generates a reachability graph from the
given specification. It then checks that the safety property is preserved
in all feasible execution paths. It can also be used to determine whether
two finite real-time processes are equivalent. As a first step, we have
implemented a reachability analyzer, which can be used to detecting
general properties, such as deadlock, livelock and possibility of an exception. We are currently working on the model checking and parallel
analysis algorithms.

The third tool, a mechanical theorem prover, uses the semantics of processes and the algebraic laws of CCSR to assist a human to develop
mathematically rigorous proofs. Unlike the reachability analyzer, this
tool can be used to reason about infinite processes using induction principles.

2.2 Extensions to CCSR

Although CCSR has been used on small toy-like problems, its applicability in practice is limited because its expressivity; actions must consume
resources and only static priorities are supported. To remedy the first
limitation, we now take the approach that systems evolve in discrete

steps, and distinguish two kinds of steps: events and actions. An event is instantaneous and does not require any resources. Events are merely named instants in time and are mainly used to achieve synchronization between processes. An action, on the other hand, takes time and is bound to resources, thus represent the work being performed by the system. For example, an action, $\chi(t, p, r)$, represents the usage of a resource, r, at a priority p, for a finite amount of time, t.

In order to arbitrate between several actions competing for the same resource, CCSR supports the notion of priority, which yields a measure of the urgency of performing an action. Priority is also used to decide between several processes attempting to synchronize with the same event. The operational semantics of CCSR is defined in terms of two extended labeled transition systems, one for events and another for actions. We define unprioritized behavior first. We then derive a notion of preemption using priorities and enhance the transition system accordingly. We also show that this preemption relation is compositional. Since the notion of locked-step parallelism is intrinsic to our model, we show how interleaving can be modelled with the addition of a "procrastinating action."

To remedy the second problem, we have extended the notion of priority to dynamic priorities. We show how dynamic priorities can be defined as a function of the history of the system and how events can be used to model some common priority schemes, such as first-in first-out and earliest deadline first. We redefine the operational semantics of CCSR to include contexts with history information. We show that, in the general case, dynamic priorities are not compositional. We give a sufficient condition that ensures compositionality and show several examples of priority schemes that are be compositional. We have investigated why it is difficult to use "local histories" to enforce compositionality.

2.3 Logic for CSR

To make the CSR/CCSR paradigm useful in practice, we are studying methods of verifying specifications by writing correctness assertions as logical formulas. Currently, we specify such assertions in CCSR as a term which exhibits the desired behavior and then show that the specification is bisimilar to the term. However, some people find it more natural to express properties as logical assertions rather than CCSR terms. If the property can only be stated as a term, such a user must convince himself that the term corresponds to the logical assertion that he had in mind. We therefore are developing a logic for CSR (LCSR) for writing partial correctness assertions and a satisfaction relation, $P \models f$, between a CCSR term P and an LCSR formula f. A subset of RTL seems to be a natural candidate for this assertion language.

3   Conclusion and Future Work

Our current goal is to develop a flexible environment in which the design of a system is specified in CSR and the requirements (i.e., desired properties) of the system are specified in a requirements specification language such as CCSR and LCSR. The design specification of a system consists of processes that define the logical constraints of functional and temporal aspects and configurations that describe the system constraints of available physical resources and underlying scheduling disciplines. The environment will also include a set of analysis tools described above. Since these tools have different strengths, they are useful in widely different situations. In particular, the integrated use of the tools allows the programmer flexibility in demonstrating that the specification is correct since different analysis tools can be used at different stages of proving that the specification is correct with respect to the requirements.

Our long-term goal is to be able to support tools based on more than one computation model, such as timed Petri net or Timed Automata, etc. In this case, the same system may be specified using different computation models, which means that there should be a way to ensure consistency between specifications written in different models. Another issue is how design and implement tools so that much of internal data structures and algorithms for analysis tools can be shared. Here, one may develop tools using an internal representation based on a very general transition system.

References

[1] R. Gerber and I. Lee, "CCSR: A Calculus for Communicating Shared Resources," *Proc. of CONCUR90, LNCS 458*, Aug 1990.

[2] R. Gerber and I. Lee, "A Resource-Based Prioritized Bisimulation for Real-Time Systems," to appear in *Information and Computation.*

[3] R. Gerber and I. Lee, "Specification and Analysis of Resource-Bound Real-Time Systems," *Proc. of REX Workshop on Real-Time: Theory in Practice*, June 1991.

[4] V. Wolfe, S.B. Davidson, and I. Lee, "RTC: Language Support for Real-Time Concurrency," *Proc. of IEEE Real-Time Systems Symposium*, Dec 1991.

A Distributed Snapshots Algorithm and its Application to Protocol Stabilization

K. SALEH[a], H. URAL[b], A. AGARWAL[a]

a Dept. of Electrical and Computer Engineering,Concordia University, Canada

b Dept. of Computer Science, University of Ottawa, Canada

Abstract

In this paper, we study the application of Chandy and Lamport's Distributed Snapshots Algorithm (DSA) [1] to compute global states of a communications protocol. In particular, we are interested in assessing the suitability of the application of DSA for protocol stabilization. We show that the protocol state obtained cannot always be used for protocol stabilization, specifically from the checkpointing or recovery viewpoint. Furthermore, we show that when a loss of coordination and synchronization occurs, DSA is not guaranteed to terminate, and therefore it sometimes fails to obtain a global protocol state. These two problems with DSA are illustrated on a protocol example.

Key Words: Communication protocols, finite state machines, global state, protocol stabilization.

1   Introduction

A distributed system is often modelled by loosely coupled processes which exchange messages over communication links. Communication links are modelled by unidirectional First-In-First-Out (FIFO) channels of unbounded capacities. A process can perform computations on local memory and variables, and can send and receive messages from other processes. A state of the distributed system (called global state or snapshot) consists of the current state of each of the component processes and the contents of the channels linking these processes. The determination of the global state of a distributed system can be useful for obtaining checkpoints for recovery and restart [2] and for deadlock detection [1, 3]. In [1] a distributed snapshot algorithm (DSA) is presented. DSA can be initiated by any process in a distributed system. Upon termination, each process will have obtained the same snapshot. DSA can be applied concurrently with an underlying distributed computation so it won't interfere with the computation itself. In this paper, the underlying distributed computation of interest to us is a communications protocol consisting of processes modelled by finite state machines. Under normal operational conditions, DSA will obtain a snapshot of the protocol and also will be able to detect a protocol deadlock. Since according to DSA, the protocol state we obtain may not be a state through which the protocol has passed, this protocol state cannot be used as a checkpoint or recovery point if later an error is detected. Moreover, DSA fails to obtain a global state once an unspecified reception error occurs in any process involved in the protocol. In fact, DSA may not terminate under certain operational conditions which cause a loss of coordination among the communicating processes. A modification to DSA is found to be necessary so that its termination is guaranteed after propagating the illegal global state. Our ultimate goal is to use the DSA to stabilize protocols. A protocol is said to be stabilizing, if starting from any illegal global state, the protocol will eventually reach a legal global state, and resumes its execution.

The rest of the paper is organized as follows. Section 2 briefly introduces the finite state machine model for communication protocols. Section 3 points out the problems with the application of DSA for protocol stabilization, and finally, Section 4 provides some concluding remarks.

2   The Communicating Finite State Machine Model

Processes in a communications protocol can be modelled by communicating finite state machines (CFSM) exchanging messages through unidirectional FIFO channels. A CFSM in a system of n CFSMs can be formally defined by the quadruple $CFSM_i = (S_i, s_0, M_{ij}, T)$, where S_i is the set of internal states of process P_i, $s_0 \in S_i$ is the initial state of P_i, M_{ij} is the set of messages sent by P_i to other processes (MS_i) and messages received by P_i from other processes (MR_i), and finally, T is

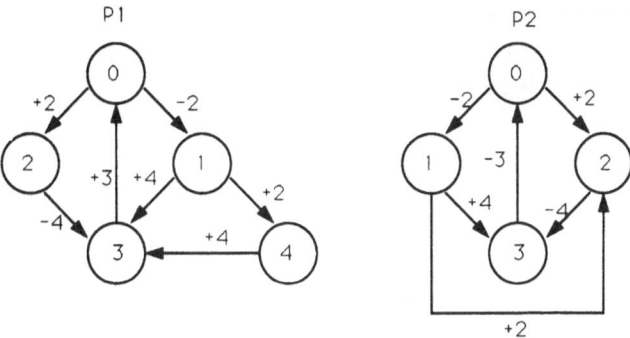

Figure 1: Protocol example with two process P1 and P2

a partial transition function: $S_i \times M_{ij} \rightarrow S_i$. We say that a message m belonging to M_{ij} is a label for a transition. A unidirectional FIFO channel C_{ij} carries messages belonging to MS_i sent from P_i to P_j. As an example, consider the protocol given in Figure 1 consisting of two processes P_1 and P_2 modeled by CFSMs. A minus (plus) sign "-" ("+") prefixing the label of a transition denotes a sending (receiving) transition. A transition at a state s of process P for message m is said to be specified if T(s, m) is defined in the CFSM modeling P. A global state of a protocol is a pair <S, C>, where S = $(s_1, s_2, ..., s_n)$, and $s_1, s_2, ..., s_n$ represent current states of processes $P_1, P_2, .., P_n$, respectively, and C = (c_{ij}, for all i $\neq$ j, and i, j $\leq$ n) represents the current contents of the channels C_{ij} linking the processes. The initial (final) global state of a protocol is a pair <S, C> in which each of the component states of S are initial (final) states in their respective processes, and all channels are empty. A global state < S, C > $\approx$ <S', C'> (<S', C'> follows <S, C>) iff there exists i, j and C_{ij} such that elements of <S', C'> can be derived from <S, C> by applying one of the following: (i) $s_{i'} = T(s_i, x)$ and $c_{ij'} = c_{ij}$ @x, x $\in$ MS_i or (ii) $s_{j'} = T(s_j, x)$ and $c_{ij} = x$@ $c_{ij'}$, x $\in$ MR_j (where @ denotes concatenation). Except for the elements affected by the application of one of the above, all other elements of <S', C'> are equal to corresponding elements of <S, C>. A global state <S, C> is said to be reachable from the initial global state <S_0, C_0>, denoted by <S_0,C_0> $\approx^*$ <S, C> iff <S_0, C_0> $\approx$ <S_1, C_1> and <S_1, C_1> $\approx^*$ <S,

C>, that is there exists an execution path consisting of an interleaving of message receptions and transmissions that takes the protocol from the initial global state $<S_0, C_0>$ to $<S, C>$. A global state $<S, C>$ is said to be legal (illegal or unsafe) if $<S_0, C_0> \approx^* <S, C>$ is (not) true.

In a protocol design consisting of the CFSM specification of each of the processes involved, there can be two types of protocol design errors [4]: i) semantic error which causes the provision of an incorrect service to the distributed protocol users and therefore affects the safety [4] of the protocol, and ii) syntactic error which ultimately causes the protocol to deadlock, and therefore affects the liveness [4] of the protocol. In the following, we further classify syntactic errors. A reception of message x at state s of P_i is said to be unspecified iff there exists a reachable global state $<S = (s_1, s_2, ..., s_n), C = (c_{ij},$ for all $i \neq j, i,j \leq n)>$ in which the following condition is true: $s = s_i$ and c_{ji} (for any $j \neq i$) $= x@Y$ for any message sequence Y, $x \in MR_i$, and both $T_i(s, x)$ and $T_i(s, y)$ are unspecified for any message $y \in MS_i$. A global non-final protocol state $<S = (s_1, s_2, ...,s_n), C = (c_{ij},$ for all $i \neq j, i,j \leq n) >$ is a deadlock state if all channels are empty and $T_i(s_i, x)$ is unspecified for any $x \in MS_i$ and any $i \leq n$, and all process states are not final states. Reachability trees have been used to analyze protocols modelled by CFSMs [5]. A reachability tree is a tree whose nodes represent legal global states, and a directed arc connecting two nodes or states $<S_1, C_1>$ and $<S_2, C_2>$ corresponds to $<S_1, C_1> \approx <S_2, C_2>$. The root of the tree corresponds to the initial global state $<S_0, C_0>$. A path in the tree corresponds to an execution sequence of interleaved receptions and transmisions and it represents the reachability of the last state in the path from the initial state of the path. The expansion of the tree from a particular node stops if one of the following conditions is satisfied: (i) the node already exists in the tree, (ii) the node corresponds to a deadlock state, (iii) an unspecified reception error is detected at the node, (iv) the node corresponds to a final global state. A reachability tree of the protocol example of Figure 1 is shown in Figure 2. It can be easily verified that the protocol is free from design errors.

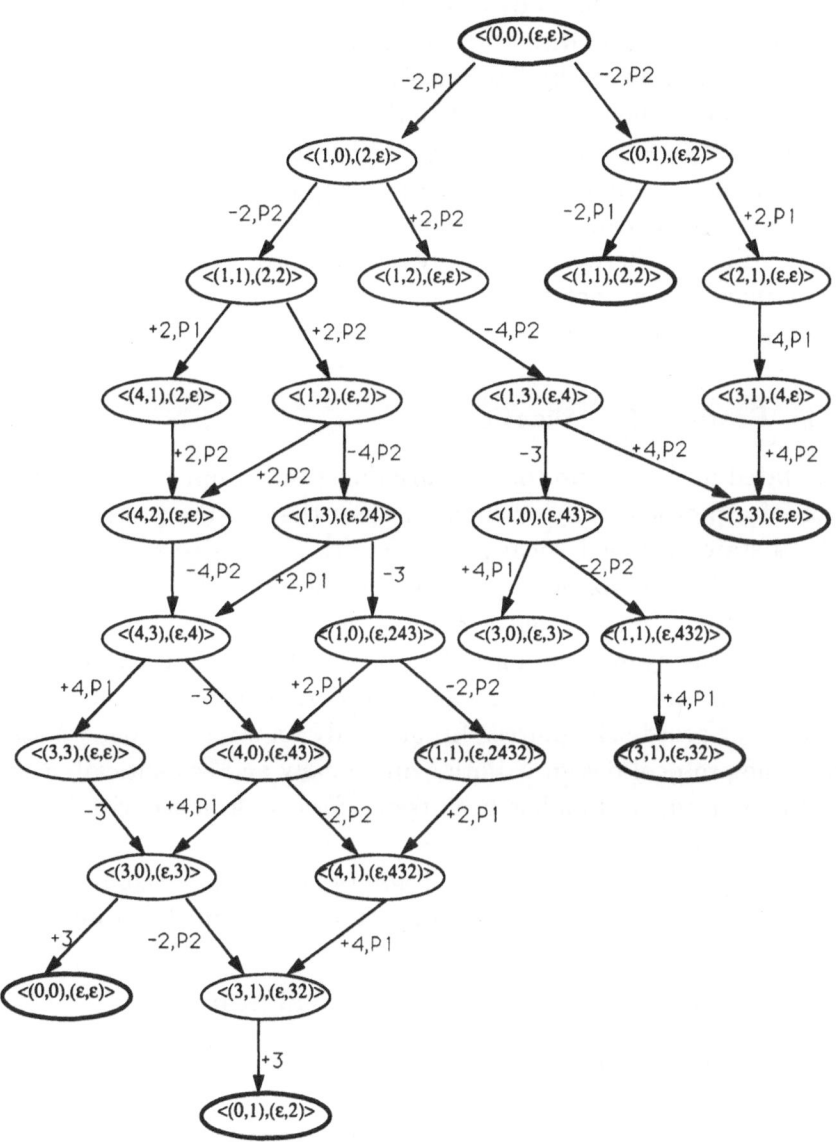

Figure 2: Reachability tree for protocol of Figure 1

3   DSA and Protocol Stabilization

DSA was first introduced by Chandy and Lamport in [1]. In this section, we point out two problems with the application of DSA to protocol stabilization. First, we briefly describe some of the issues related to protocol stabilization. Then, we show that the snapshot obtained by DSA may or may not be a state through which the protocol have passed. Therefore, this state cannot be used as a recovery point for backward recovery once a protocol error is detected later. Finally, we show that after a loss of coordination between the processes of a protocol, the application of DSA may or may not terminate, therefore, it is not guaranteed that DSA will return any global state.

3.1   Protocol Stabilization

As stated in [6], one protocol failure that can disrupt the proper execution of a protocol is a coordination loss, that is, the protocol is in an illegal state, although local process states may be correct. A protocol is said to be stabilizing, if starting from any illegal state, the protocol will eventually reach a legal state, and resumes its normal execution. There are many reasons for a loss of coordination in a protocol such as: incorrect initializations, memory overwrites, transmission errors and process or processor failures. In our study on stabilization, we assume that the protocol design is both syntactically and semantically correct, and its implementation has been tested for conformance with its design specification.

In the following, we assess the suitability of DSA for stabilization after a loss of synchronization and coordination occurs in the protocol.

3.2   Usefulness of the Obtained Global State

According to [1], if DSA is started in state S_0 (of the protocol), it terminates in some state S_1 and returns as its result a snapshot state S_s. This state S_s, although it may not have actually occurred, has the following properties: 1) it is possible for the protocol to reach S_s from S_0, and 2) it is possible to reach S_1 from S_s. To illustrate the above, let us consider the protocol example of Figure 1 and its reachability tree in Figure 2. Assume that DSA is initiated at state $<(1,2),(e,e)>$ by P_1, so the state recorded for P_1 is 1. After recording its state, P_1 sends a marker along channel C_{12}. Now the system may go to state $<(1,3),(e,4)>$, then $<(3,3),(e,e)>$ and then $<(3,0),(e,3)>$ while the marker is still in transit, and the marker is received by P_2 when the system is in state $<(3,0),(e,3)>$. On receiving the marker, P_2 records its state as 0 and the content of C_{21} as the sequence of messages 4 then 3. The recorded global state is $<(1,0),(e,43)>$. The state obtained by DSA is a legal one since it is included in the protocol reachability tree. However, although it is a legal state, it is not a state that actually occured during the execution of the protocol. Figure 3 shows the part of the reachability tree which includes the protocol execution path and the recorded state. Because of the above observations on DSA, the recorded state cannot be used as a checkpoint for the backward recoverability and stabilization of protocols. Resuming the execution of the protocol from such state may lead to undesirable effects such as providing the wrong service to the protocol users. However, the snapshot obtained by DSA would be enough to detect deadlock states, since if S_s is a deadlock state so is S_1 and if S_s is not a deadlock state so is S_0.

3.3   Handling of Unspecified Receptions

In the protocol example, an incorrect initialization of P_1 and P_2 to states 4 and 1, respectively, results in a deadlock global state. An application of DSA will successfully detect such operational error [3]. However, in the same protocol, if after P_1 and P_2 exchange message 2, the state of P_1 becomes 3 because of a memory error and the state of P_2 becomes 3 after sending 4 to P_1. The protocol state $<(3,3), (e,4)>$ is an illegal state since it corresponds to an unspecified reception error. Furthermore, DSA will not terminate, even if it was initiated by any of the two processes. Because we are dealing with FIFO channels, P_1 will never receive the marker m originating from P_2, and therefore, no protocol state will be obtained. Some necessary and sufficient modifications to DSA to allow it to terminate after propagating the illegal global state are provided in [7]. These modifications are based on the use of a special marker m' transmitted by the process which encounters an unspecified reception. An alternative implementation-oriented strategy based on a priority queue for an out-of-order reception of unspecified messages can also be used. According to [1], to ensure the termination of DSA, each

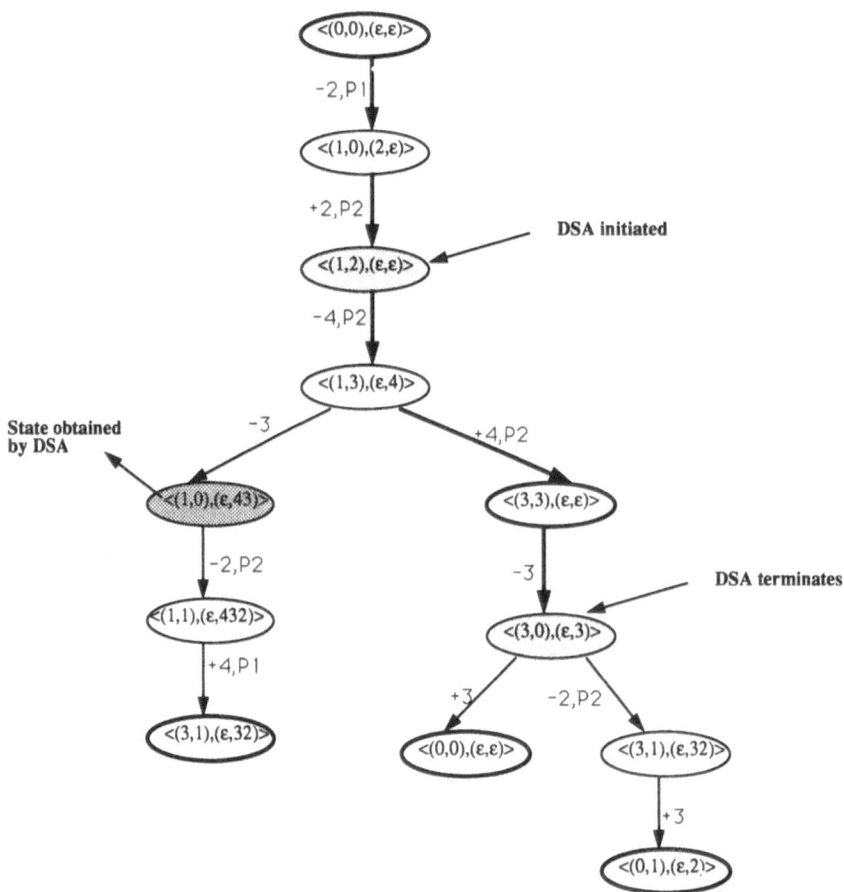

Figure 3: Part of the reachability tree which shows the computed state and execution path

process must ensure that (1) no marker remains forever in any of its input channels, and (2) its local state is recorded within finite time after initiation of DSA. Condition (1) is only concerned with the finiteness of time between the arrival of marker m at the head of the FIFO queue and its actual acceptance or reception by the process. However, this condition does not deal with the arrival of m when it is preceeded by one or more messages and in which case the reception of the message at the head of the queue is not specified in the process. The inability of DSA to handle such abnormal condition affects its termination, and therefore its suitability for application to stabilize communication protocols.

4 Conclusion

In this paper, we have studied Chandy and Lamport's DSA in order to assess its suitability for the design of stabilizing communications protocols. We have shown that, DSA, when applied to a communications protocol after a loss of coordination occurs, may not terminate. Moreover, under normal protocol execution condition, the application of DSA will return a protocol state which may or may not be a state through which the protocol passed after the initiation of DSA and before its termination. Therefore, because of this specific feature of the algorithm, the obtained snapshot cannot be used as a recovery point if later an illegal protocol state is obtained, and a backward recovery [8] is needed. Currently, we are working on the design of efficient distributed algorithms for obtaining global states in distributed systems which can be used for detecting protocol execution problems and for optimal rollback recovery [9] from those execution problems.

References

[1] K. Chandy and L. Lamport, "Distributed snapshots: determining global states of distributed systems," *ACM Trans. on Comp. Sys.*, vol. 3, no. 1, pp. 63–75, 1985.

[2] C. Morgan, "Global and logical time in distributed algorithms," *Information Processing Letters*, vol. 20, pp. 189–194, 1985.

[3] M. Raynal, *Networks and Distributed Computation*. The MIT Press, 1988.

[4] R. Probert and K. Saleh, "Synthesis of communications protocols: survey and assessment," *IEEE Trans. on Computers*, vol. 40, no. 4, pp. 468–476, 1991.

[5] G. Bochmann, "Finite state description of communication protocols," *Computer Networks*, vol. 2, no. 4/5, pp. 361–372, 1978.

[6] M. Gouda and N. Multari, "Stabilizing communication protocols," *IEEE Trans. on Computers*, vol. 40, no. 4, pp. 448–458, 1991.

[7] H. U. K. Saleh and A. Agarwal, "A modified distributed snapshots algorithm for stabilizing protocols," *submitted for publication.*

[8] P. Lee and T. Anderson, *Fault tolerance: Principles and Practice.* Springer Verlag, 1990.

[9] K. Saleh and A. Agarwal, "Efficient and fault tolerant checkpointing procedures for distributed systems," *to appear in Proceedings of the Intern. Phoenix Conf. on Computers and Communications*, 1993.

Protocol Validation Tool and Its Applicability to Responsive Protocols

H. SAITO, T. HASEGAWA

KDD R & D Laboratories

2-1-15 Ohara, Kamifukuoka-shi

Saitama, Japan

Abstract

A protocol specification, based on which a communication system is built, should be designed without an error because a late error detection makes system correction unacceptably expensive. It is, however, becoming more and more difficult to design error-free protocol specifications for the modern complex communication systems. Such a situation causes a great demand for tools that automatically verifies properties required for protocols. One of the required properties is freedom of logic errors such as undefined reception or deadlock. Another property which is essential to reliable systems is responsiveness, i.e., a fault-tolerant and real-time property. In this paper, we introduce our protocol validation tool that verifies freedom of logic errors using the acyclic expansion algorithm. Then we discuss its applicability to the verification of responsiveness.

Key Words: Protocol validation, acyclic expansion algorithm, responsive protocol, fault-tolerance, real-time.

1   Introduction

In the development of a communication system, a protocol specification which prescribes the manner of message exchange among processes is designed. An error in the protocol specification will remain in both the detailed specification and the system implemented according to the specification. Eliminating all errors in a protocol specification is desirable because a late error detection makes system correction unacceptably expensive. It is, however, becoming more and more difficult to make an error-free protocol specification or to detect all protocol errors in a specification, since a communication system is now getting more and more large-scaled and complicated. Such a situation causes a great demand for tools that automatically verifies the required properties of a protocol specification. For example, freedom of logic errors. i.e., undefined reception or deadlock, must be verified, which is called *protocol validation*. The verification of a fault-tolerant and real-time property, which is called *responsiveness*, is also essential to make the system reliable. Among the above properties, the freedom of logic errors is verified using the *acyclic expansion algorithm* [1]. The responsiveness can also be verified by extending the algorithm [2].

This paper will introduce a protocol validation tool we have developed on the basis of the acyclic expansion algorithm. The tool has the following features, and supports the efficient protocol design.

1. high-speed and exhaustive validation based on the acyclic expansion algorithm;

2. graphical output which makes it easy to understand the cause of errors; and,

3. functions which enables stepwise validation and partial validation.

Moreover, we will discuss the applicability of the tool to the verification of responsiveness.

This paper is organized as follows. First, the concept of protocol validation is explained in Section 2. The features of the protocol validation tool are described in Section 3. The validation method on which the tool is based is introduced in Section 4. The configuration and input and output examples of the tool are shown in Section 5 and Section 6, respectively. The industrial application of the tool is reported in Section 7. Then, we discuss its applicability to verification of responsiveness in Section 8.

2   Protocol Validation

A protocol is realized by processes which exchange messages and channels through which the messages are transmitted. Here we model a process as a finite-state machine (FSM) which changes its state every time it receives or transmits a message and a channel as a full-duplex, error-free, FIFO queue. Then, a protocol specification is described as a set of state transitions for the processes. For example, Figure 1 shows a specification of a three-process protocol described as state transition diagrams. In the figure, a circle denotes a state of a process, and an arrow denotes a state transition. A label attached to an arrow denotes an event which invokes the associated state transition, where plus sign identifies the reception of a message, minus sign its transmission. In the example of the figure, process 1 when in state 1 can enter a new state 2 by transmitting a message cr.

Since a protocol specification is described as the behavior of each process, there is a high possibility that a description of necessary message transmissions or receptions are missing, or that unnecessary message transmissions or receptions are described in the specification. Especially, missing receptions are prone to occur, since a transmitted message has possibility of being received in more than one process state. Besides, a missing transmission might cause an error called *deadlock*. Such errors are significant design errors which lead to fatal system failures when implemented. Typical errors are as follows:

- Undefined reception

  no message reception is described in a receiving process, while the message is transmitted to the process

- Unexecutable reception/transmission

  reception/transmission of a message which is described but never executed

- Deadlock

  a system state which allows no more execution of events since all processes are at the states where no transmission is specified

- Channel overflow

  a state of a channel in which messages are accumulated exceeding the specified channel capacity

Such errors common to all protocol specifications are called *logic errors*, and detection of logic errors in a protocol specification is called *protocol validation* in this paper.

When a logic error is detected by protocol validation, the protocol specification is to be modified. The modified version of specification might then have a new error which was introduced when modified. Therefore, protocol validation and modification should be executed iteratively. The protocol validation tool not only detects logic errors in a protocol specification but also provides an environment which supports the protocol design phase including protocol validation and modification.

3 Features of the Tool

In order to make the validation and modification work more efficient, the protocol validation tool has the following features.

1. High-speed and exhaustive validation

   The tool performs high-speed and exhaustive protocol validation using the acyclic expansion algorithm. The details of the method is described in Section 4.

2. Graphical outputs that support error correction

   For correcting an error detected by the tool, the sequence of message transmissions and receptions is useful rather than the state at which the error was detected. The tool generates message sequence charts which show the sequences leading to detected error states.

3. Stepwise and partial validation

   The tool has an interactive mode which allows a user to control the validation by typing commands. For example, it is possible to stop validation when an error is detected. In a large protocol specification, an error is apt to cause some other errors. A user can efficiently correct the error at the moment of detection, avoiding the meaningless continuation of validation.

   Also, the scope of protocol validation can be easily specified (See Section 6). This enables partial validation of a specification, which is particularly convenient for a large protocol specification.

4 Acyclic Expansion Algorithm

For protocol validation, it is necessary to obtain all executable state transitions, i.e., message transmissions and receptions that have possibility of being executed when the system runs.

In most of the traditional protocol validation methods [3], global states reachable by executable state transitions are exhaustively enumerated and then the existence of logic errors are checked at every global state. Here, a *global state* is a state of the system consisting of the states of all processes and channels. With the increase of the number of processes, the number of global states to be enumerated increases exponentially, which is so-called *state explosion* problem.

On the other hand, the validation tool employs a different protocol validation method called acyclic expansion algorithm [1]. The method generates a state transition graph consisting of executable transitions for each process, checking the correspondence of a message between its transmission and reception. During construction of the executable state transition graph, or *expansion*, such errors as undefined receptions are automatically detected. The method is considered efficient for the following reasons: Only the processes transmitting and receiving a message are expanded at one time; and an executable state transition graph is obtained for each process graph instead of a large graph representing the whole system behavior. Logic errors are all detected by checking the executable state transitions.

Figure 2 shows the executable state transition graphs obtained by applying the acyclic expansion algorithm to the protocol specification in Figure 1. A state in the specification appears more than once as reachable states in the graph. Similarly, a state transition appears several times as executable transitions. To identify each reachable state and executable transition, we attach a sequential number to its original name. For example, the state 6 reached by transmitting message cq from state 1 in process 1 is named as 6.1 and distinguished from state 6.0 which is reached by transmitting cq from state 5. The reception of cc.0, which is denoted for a dotted line in the process 1, is an example of the undefined reception error. The transition +cq from state 4 in process 2 is defined in the specification of Figure 1, but not included in the obtained executable state transition graph in Figure 2. This is detected as an

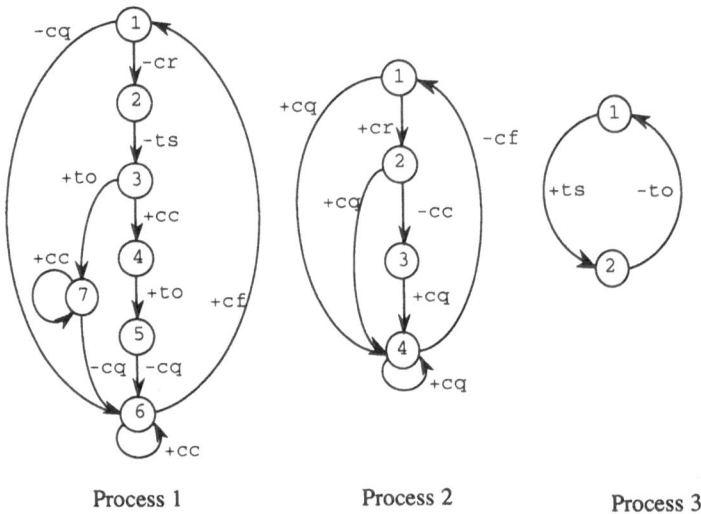

Figure 1: Example of an FSM-based Protocol Specification

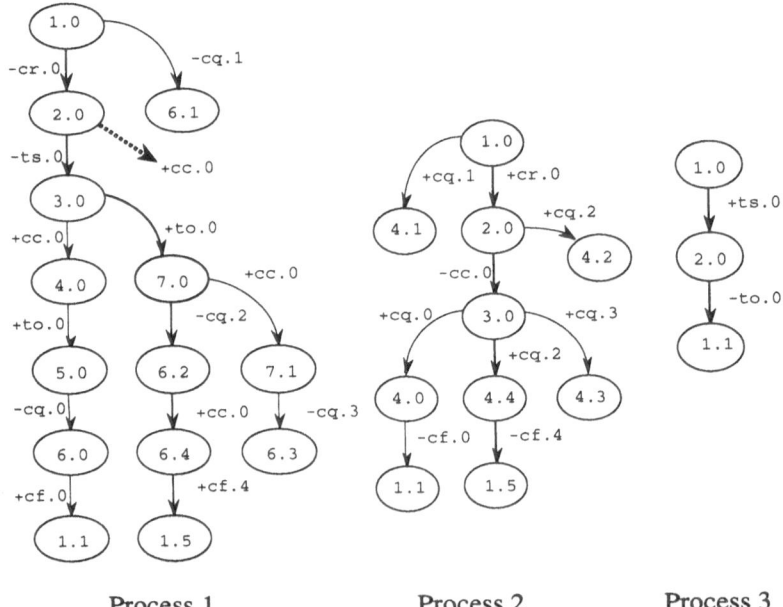

Figure 2: Example of Executable State Transition Graphs

unexecutable transition.

In order to attain more efficient validation, the algorithm stops the expansion of a sequence when one of the following is predicted.

(a) Further expansion will be the iteration of a sequence.

(b) Further expansion will be equivalent to the expansion of another sequence.

For example, the sequences following the states 1.1 in process 1 and the state 4.1 in process 2 are stopped because (a) and (b) are predicted, respectively.

5   Configuration of the Tool

The protocol validation tool has been written in C language and is runnable on any UNIX machine, while it is now running on a SPARC-station. Also, neither a special printer, a graphical terminal, nor a window system is indispensable for the tool, since all the input and output files are made of characters. (The tool can also run on a window system [4].)

The configuration of the tool and the data flow among the modules are shown in Figure 3. In the figure, a software module is denoted by a rectangular, an input/output data file and an intermediate data file are denoted by an ellipse of a thick line and an ellipse of a fine line, respectively. The functions of each module are outlined as follows.

1. Preprocess Module

   This module transforms the input protocol specification into the form which is easier for the validation module to interpret. In the protocol specification file, a protocol is described as state transitions of each process (See Section 6). During the transformation, lists of messages, states, etc. are created and output into the preprocess report file. Simultaneously, simple errors such as transmission loops or mistakes regarding the input format are reported.

2. Validation Module

   This is the main module of the tool which expands an input protocol specification using the acyclic expansion algorithm and detects
   errors such as undefined receptions. The result is stored as a binary
   file to spare the disk space. An intermediate result of the protocol
   validation can also be stored as a file. In such a case, the validation
   can be restarted later using the intermediate result file.

3. Report Generation Module

   This module generates a validation report out of the validation result
   file. Auxiliary information such as the expanded transitions can be
   output into the report.

4. Sequence Chart Generation Module

   This module generates sequence charts out of the validation result
   file. The sequence charts which lead to the specified states can be
   generated as well as the sequence charts which lead to the detected
   errors.

6   Input and Output of the Tool

Figure 4 shows an example of input to the validation tool. It is part of
the specification of ISDN User Part, which deals with call control, e.g.,
establishment or release of a call, in the ISDN.
In the protocol specification, an initial state and end states are specified
prior to state transitions in each process. The expansion of the process
is started from the state specified as initial state, and the expansion
of the sequence following the states specified as end states are not executed. Using the initial state and end states, the scope of validation
can be controlled without changing the part of state transitions. In the
example, state IDLE is specified as an initial state, state ICCAN as an
end state.
A state transition in the protocol specification consists of a state where
the transition occurs, a state which the process enters by the transition,
and a condition for the transition, where the condition is normally reception or transmission of a message. For example, the 5th. line in Figure

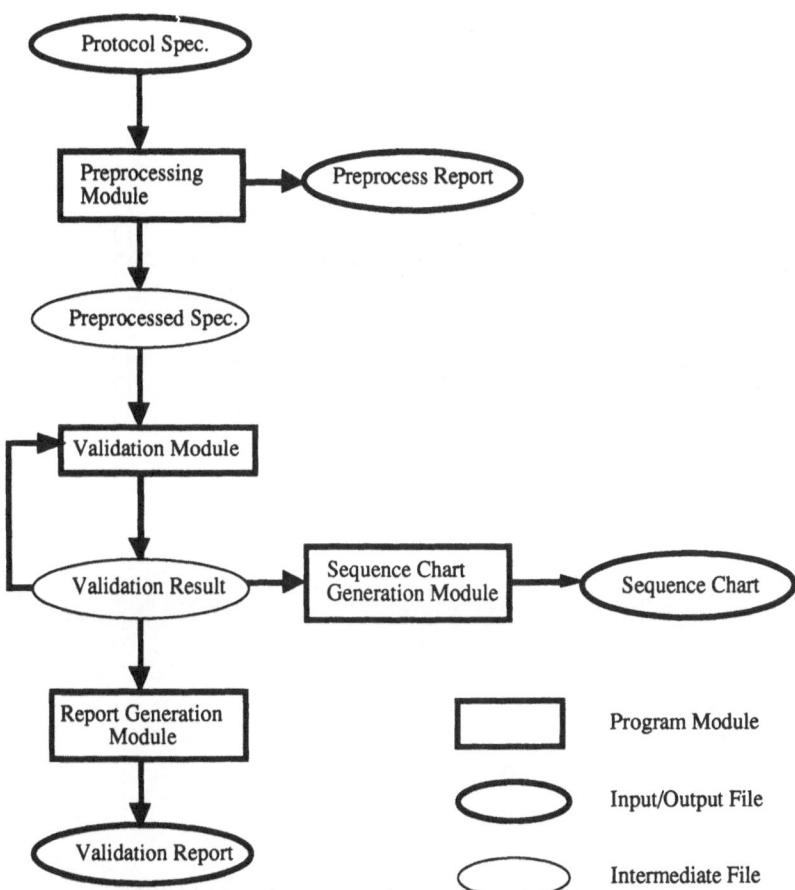

Figure 3: Configuration of the Validation Tool and Data Flow

```
Protocol ISDN_UP(
Process CPCI (
               Init  IDLE
               END   ((ICCAN,1))
        (IDLE     ,_S01     , (+IAM   ,MDSC    ))
        (_S01     ,_S06     , (-SCPCI ,MDSC    ))
        (_S06     ,WFACM    , (-SUID  ,CC      ))
        (WFACM    ,_S35     , (+ALRQ  ,CC      ))
        (_S35     ,WFANM    , (-ACM   ,MSDC    ))
        (WFANM    ,_S53     , (+SURP  ,CC      ))
        (_S53     ,ICCAN    , (-ANM   ,MSDC    ))
        (ICCAN    ,_S67     , (+REL   ,MDSC    ))
                       .
                       .
                       .

)
Process CPCO (
                       .
                       .
                       .

)
       .
       .
)
```

Figure 4: Input Format of Protocol Specification

4 indicates that a transition of CPCI from state IDLE to state _S01 occurs when message IAM is received from process MDSC. The next line indicates that the transition from state _S01 to state _S06 occurs by transmitting message SCPC to process MDSC.

A protocol specification in this form should be prepared with an editor, otherwise it can also be obtained by automatic conversion from a specification written in SDL(Specification and Description Language) [5], allowing the SDL specification to be validated [6].

As an example of output, a sequence chart leading to a detected undefined reception error is shown in Figure 5. The error occurs at the state WFIRC of the process CPCI when the message ALRQ arrives from the process CC, because the reception of ALRQ is not defined in the specification. However, the real cause of the error is not the lack of the reception but the transmission of RLD mistakenly specified following the transmission of SUID. Thus, it is necessary to analyze the sequence of message exchange for determining the real cause of a detected error, and sequence charts are quite useful for this purpose. Not only erroneous sequences, but also sequences leading to a state specified by a user are obtained. Such sequence charts can be used to confirm that intended sequences are represented by the specification.

7   Industrial Application of the Tool

The validation tool has been used to validate a protocol specification for the mobile satellite communications system of INMARSAT (International Maritime Satellite Organization). The system consists of Land Earth Station, Network Coordination Station and Mobile Earth Station, and exchanges messages among them for a call control. The validated protocol specification has 27 processes and 185 process states. The validation and modification has been iterated 38 times and totally 302 undefined reception errors have been detected and corrected.

The validation took approximately 25 minutes each time, and 16 hours as a whole. The modification took 1 to 2 hours each time, and about 60 hours as a whole. Total time spent is about 76 hours. For a comparison purpose, we here assume that a man can check the executability of ten transitions per minute. Then, the time required for the same work would amount to 1,053 hours if it is done manually, since totally 632,000

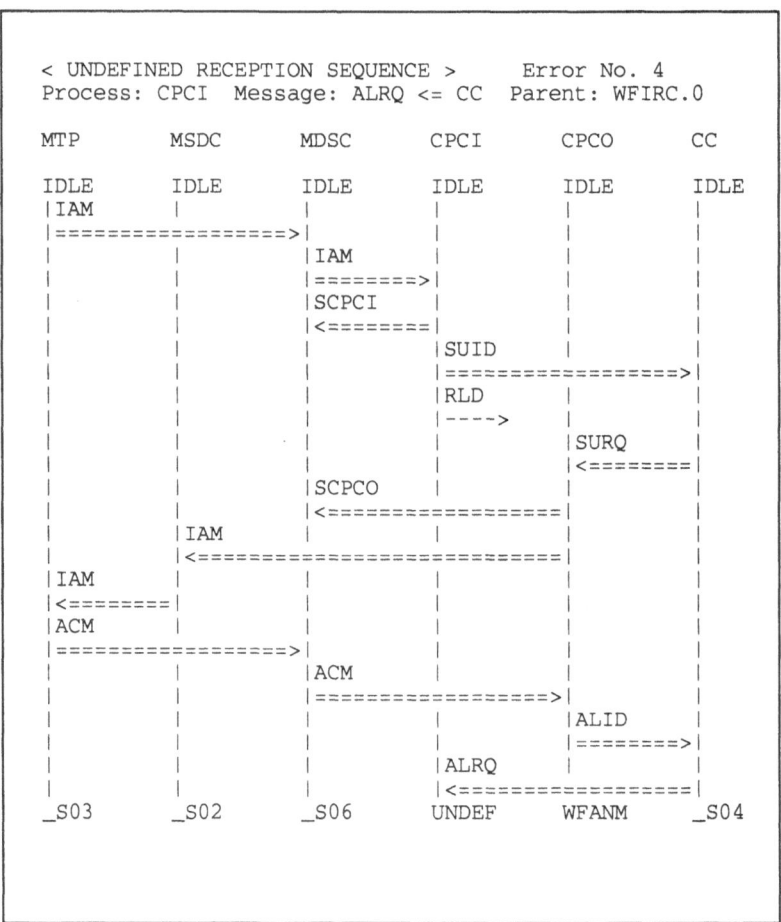

Figure 5: Sequence Chart Leading to an Undefined Reception Error

transitions were checked by the tool during the validation of 38 times. This simulation yields a result that the use of the tool has brought about 14 times more efficient validation work.

8   Applicability to Responsive Protocols

Responsiveness is defined as a fault-tolerant and real-time property [7]. For developing a reliable system, it is important to take into account the responsiveness from the design phase, and to verify the responsiveness of the protocol. It has been argued that the acyclic expansion algorithm, on which the validation tool is based, is extended such that the responsiveness can be verified [2]. This section presents how the tool can be applied to the verification of responsiveness, i.e., the fault-tolerant property and real-time property.

8.1   Fault-Tolerant Property

If a protocol is fault-tolerant, it never ends with an abnormal state, but eventually reverts to a normal state [8]. Let us define a normal global state as a global state which is a set of normal process states, and an abnormal global state as a global state which includes at least one abnormal process state. Whether each process state is normal or abnormal should be specified in a protocol specification in any way. Then the property is verified by checking the executable state transition graphs so as to whether the following requirement is satisfied.

For every process,

> for every abnormal process state,
> the state eventually reaches a normal process state which
> satisfies the following condition.

Condition: A stable global state which the process state belongs to is a normal global state, where *stable global state* is a global state with all the channels empty.

To embed the checking function into the tool, a new program module may be developed. Otherwise, the formula evaluation module which is described in [9] may be utilized. In this case, the above requirement should be input to the module as branching-time temporal logic formulas.

8.2 Real-Time Property

Real-time systems [10] have the requirement of completing tasks within a specific time. In order to verify the real-time property, it is necessary to calculate the time for state transition sequences that correspond to time-critical tasks. This can be done by placing each state transition of each process on the time axis, considering the *precedence relation* [2] between state transitions of different processes.

For the verification of the real-time property, the following functions should be realized:

1. identify state transition sequences that correspond to time-critical tasks;

2. calculate the time for those sequences considering the precedence relation; and,

3. verify that the execution time for the sequences does not exceed the given time even in the worst case.

While it is assumed in the verification method in [2] that all transitions take one unit time, more precise verification may be possible by assigning different time to individual transitions in a protocol specification.

9 Conclusion

This paper has introduced the protocol validation tool based on the acyclic expansion algorithm. The tool performs protocol validation exhaustively and at a high speed. With its graphical outputs, and useful functions, the tool supports the efficient validation and modification of protocol specifications. Its effectiveness has been shown by the protocol validation of the INMARSAT mobile satellite communications system. Moreover, its applicability to responsive protocols has been discussed, and some necessary functions of the tool for verifying responsiveness have been shown. We are intended to continue the study on the more detail functions needed for the extension of the tool.

Acknowledgements

The authors would like to express their sincere thanks to Dr. K. Ono, Dr. Y. Urano and Mr. K. Konishi of R&D Laboratories of Kokusai Denshin Denwa Co., Ltd.(KDD) for their general guidance and encouragements. They also greatly appreciate Dr. Y. Kakuda of Osaka University for his useful suggestions on this study.

References

[1] Y. Kakuda, Y. Wakahara, and M. Norigoe, "An acyclic expansion algorithm for fast protocol validation," *IEEE Trans. Software Eng.*, vol. SE-14, no. 8, pp. 1059–1070, Aug. 1988.

[2] Y. Kakuda and T. Kikuno, "Verification of responsiveness for communication protocols," *Rep. of Tech. Group, IEICE of Japan, FTS91-57*, Dec. 1991.

[3] M. T. Liu, "Protocol engineering," *ADVANCE IN COMPUTERS*, vol. 29, pp. 110–113, New York: Academic, 1989.

[4] H. Saito and Y. Kakuda, "Protocol validation system based on the acyclic expansion algorithm," *Rep. of Tech. Group, IEICE of Japan, SSE90-18*, May 1990. in Japanese.

[5] "CCITT Recommendation Z.100, Melbourne," Nov. 1988.

[6] A.Ito, E. Utsunomiya, F. Nitta, Y. Kakuda, and H. Saito, "Evaluation of tools in ESCORT," *SDL'91:Evolving Methods*, pp. 437–448, North-Holland, 1991.

[7] M. Malek, "Responsive systems (A challenge for the nineties)," *Microprocessing and Microprogramming*, vol. 30, pp. 9–16, North-Holland, Aug. 1990.

[8] E.W.Dijkstra, "Self-stabilizing systems in spite of distributed control," *Communications of the ACM*, vol. 17, no. 11, pp. 643–644, Nov. 1974.

[9] H. Saito, T. Hasegawa, and Y. Kakuda, "Protocol verification system for SDL specifications based on acyclic expansion algorithm and temporal logic," *4th Int'l Conf. on Formal Description Techniques (FORTE'91)*, pp. 513–528, Nov. 1991.

[10] J. A. Stankovic, "Misconceptions about real-time computing: A serious problem for next-generation systems," *Computer*, vol. 21, no. 10, pp. 10–19, 1988.

Keynote Address

The Concepts and Technologies of Dependable and Real-time Computer Systems for Shinkansen Train Control

A. HACHIGA

Research Division for Information and Control Systems

Railway Technical Research Institute

Kokubunjishi,Tokyo 185, Japan

Abstract

The Shinkansen is known to be a rapid, mass and reliable transport in Japan. This paper describes a dependable computer system which manages train operation and controls train route. Design requirements of safety and fault-tolerance are described. Railway-specific concepts to realize fail-safe computer are explained, together with technologies to implement a fail-safe and fault-tolerant computer system in railway. Field data and operation records over 15 years in the Shinkansen are illustrated in figures.

Key Words: Train traffic control, Fail-safe computer, Fault-tolerant, Diagnosis in dual computers

1 Introduction

Today, the average number of daily passengers carried by the Shinkansen network exceeds 700,000. At peak times trains run at the intervals of 3.5 minutes. The safety that the Shinkansen has operated without a single fatal accident to passengers in the 28 years since its 1964 inauguration and the reliability that the average delay per train is less than 1 minute qualify it as the most important public transport system in Japan.

The development of a real-time computer system started around 1964 in Japan and the studies of computer applications to seat reservation and yard automation began about the same time in the Japanese National

Railways(JNR) as well. Instead of using the computer the Shinkansen adopted wire logic ATC (Automatic Train Control) and CTC (Centralized Traffic Control) as its train control systems. The reason seemed that the computer at that time was not reliable enough to be used in daily train operations. It was ten years later that the Shinkansen applied computers to control trains for the first time. This computer system is called COMTRAC (COMputer aided TRAffic Control system) and today it is almost impossible to imagine train operation in the Shinkansen without COMTRAC. The basic functions of COMTRAC are to generate train schedule, to help the dispatcher regulate traffic disturbances, to control train routes and to provide passengers in station with such traffic information as delay or departure time. And the use of computer for CTC has also materialized a few years later.

In this paper we present COMTRAC and CTC in the Shinkansen, putting emphasis on both design and record of the fault-tolerant computer systems. And we also touch on the concept and implementation of safety in railway. The remainder of the paper is organized as follows: Section two outlines the Shinkansen network; Section three summarizes the requirements of train control in the Shinkansen; Section four introduces COMTRAC; Section five illustrates the dependable technologies for train control in railway; Section six gives the record of field data of the computer systems in the Shinkansen. Conclusions appear in Section seven.

2   Outline of the Shinkansen Network [1]

The Shinkansen network of today in Japan is shown in Figure 1. Beginning with the Tokaido line in 1964, Shinkansen lines have been constructed one after another to form a nationwide network. The changes in traffic volume and train speed of the Shinkansen to date are shown in Figure 2. The decrease of passengers in the second half of 1970s was due to the oil crisis and fare increases. After the privatization of JNR the traffic volume is increasing satisfactorily. Today Japan Railway companies are making efforts in raising the maximum train speed.

The responsiveness of the Shinkansen train control system is observed on the charts in Figures 3 and 4. In Japan the numbers of people who were killed or injured on board per thousand million man kilometre are

Figure 1: Shinkansen Network in Japan

represented by the barcharts in Figure 3. It is clear that the Shinkansen is superior to the other modes of transport as far as the safety is concerned. The average delay per train is less than 1 minute as shown in Figure 4. This is the reason why most Japanese people choose the Shinkansen rather than the air line for their trip of about three hours.

3 Requirements of Train Control in the Shinkansen

Dangers in train operation are collision and derailment. Rear-end, head-on and side-on collisions may occur between trains. Train may collide with car at level crossing. Fortunately there is no level crossing on the Shinkansen. Being free from collision with car increases considerably the safety of the Shinkansen. Derailment is caused by excess speed, wrong switch position and so on. The purpose of train control is to carry the passengers and goods to their destinations, while preventing them from encountering these dangers. Train speed control and train route control are needed in order to let it run safely. In modern railway we can control train route by operating interlocking device through CTC and

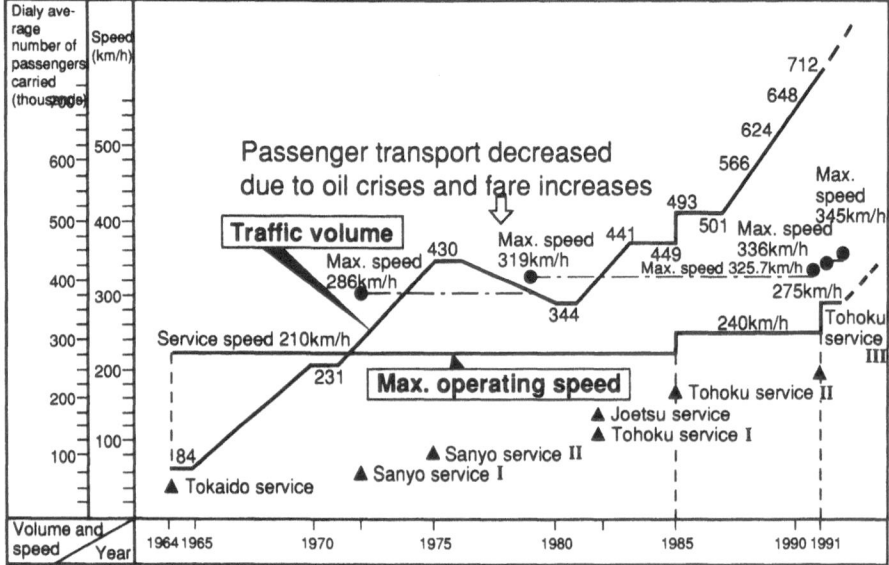

Figure 2: Traffic Volume and Train Speed in the Shinkansen

can control train speed by ATC. These special hardwares are designed to meet the fail-safe concept of railway.

The concept of route interlocking in station is depicted in Figure 5. A route is a section of tracks where train is allowed to run according to the indication of a signal. We can find ten routes on the track layout in Figure 5, some of which share parts of tracks with others. When one route is occupied, some of the others, which share some part of tracks with this one, should be locked and the rest are left free. Interlocking device is responsible for the safe control of train route in station by taking the routes which share tracks into consideration.

The concept of ATC is depicted in Figure 6. Tracks of the Shinkansen are divided into a lot of sections with 3 km length each. Each section has an insulating joint on both ends to form an electric circuit. The ground component of ATC controls the frequency of electric current flowing in this circuit. This frequency corresponds to the upper limit of speed at which train is allowed to run on the section. On-board component of ATC picks up the frequency from the track circuit to get the reference

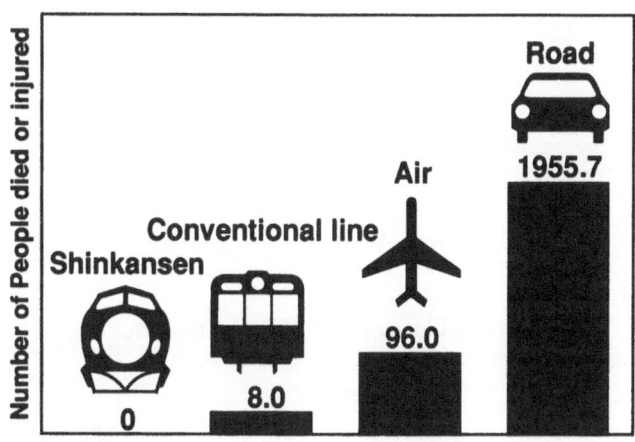

Figure 3: Safety Comparison between Major Modes of Transport

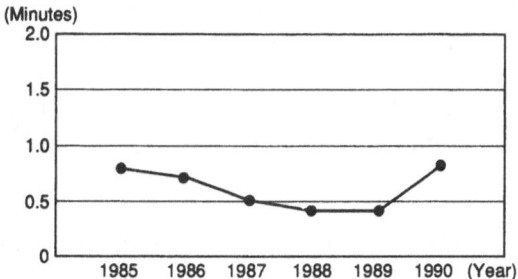

Figure 4: Average Delay per Train

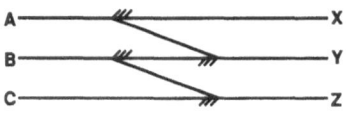

Setting	Allowed	Forbidden
A → Y	B → Z	A → X
	Z → B	X → A
	C → Z	B → Y
	Z → C	Y → A
		Y → B

Figure 5: Concept of Route Interlocking

speed. If the actual value of the train speed exceeds the reference value, ATC brakes the train to run safely. Train speed on each section is originally restricted by physical, geological and occasional conditions. Consequently the speed limit on one section is equal to the smaller of the two; the speed limit related on the next section and the speed limit on the section. Trains can reduce their speed to the reference value within the section.

Requirements on fault-tolerance of the computer systems for train control and data processing in the Shinkansen are as follows;

(a) The operation time of the Shinkansen is supposed to be 18 hours a day from six in the morning to twelve midnight.

(b) System down of an hour per year seems to be a realistic target if we consider the influence of system down by failure of facility in the past.

(c) MTTR is preferable to MTBF because the amount of passengers would not admit few but long discontinuations of train operation on the Shinkansen.

(d) The computer system for train control should have higher availability than that for data processing because the latter could send control data to the former in advance.

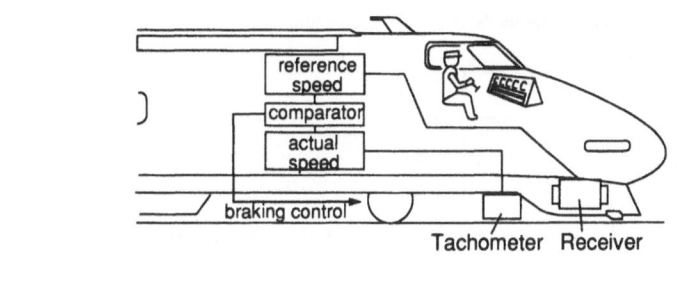

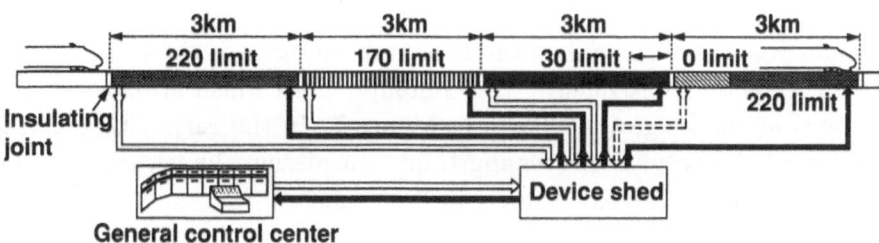

Figure 6: Concept of Automatic Train Control

(e) We would like to expand the software concurrently with train operation because most of the engineers prefer daytime to nighttime for maintenance and expansion of the system.

Considering these conditions, we have concluded that 99.99% and 99.999% are the target availabilities of data processing and train control systems respectively.

Before discussing the real-time feature of COMTRAC system, we should talk about CTC. Modern railway has a CTC system to control remotely signals and switches along tracks. A CTC is essentially a dedicated highly reliable transmission line with polling function. The central unit sends control data to signals and switches and receives the indications of signals, switches and track circuits periodically. To date the computer control of train route has necessarily made use of CTC as the transmission line between the computer and the wayside equipment. The Shinkansen is not an exception.

The polling period of CTC is related directly to the real-time feature of COMTRAC. It is 3 seconds today. Therefore the response time of train control system is a multiple of 3 seconds. On the other hand in JR

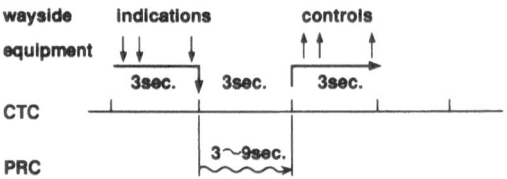

Figure 7: Response Time of Train Control Computer

(Japan Railway Companies) the time unit to manage the train should be 15 seconds. Consequently the computer must finish processing in 9 seconds at the longest as shown in Figure 7. In the early COMTRAC we accepted 6 seconds processing time, considering the performance of computer. Today we cut it down to 3 seconds.

4   Computer Aided Traffic Control System in the Shinkansen[2,3]

4.1   The Background of Traffic Control by Computer

During the first decade of the Shinkansen, the dispatcher in the General Control Center controlled train route by operating remotely the interlocking device in station through CTC. In those days two kinds of trains,i.e., express and super express, ran alternately between Tokyo and Osaka every half an hour. Express stops at every station and the stations where super express stops were fixed. As a result the dispatcher could make easily the remote control of inbound/outbound routes by himself. However the extension of the Shinkansen has changed the situation. The number of trains per hour has increased remarkably, besides the stations where super express stops are variable depending on the individual train. Therefore it seemed difficult for the dispatcher to control train route without a mistake. Finally we have decided to introduce COMTRAC into the Shinkansen.

4.2 Configuration of Computer Systems

Figure 8 shows the configuration of COMTRAC computer systems. Broadly speaking one third on the left hand side corresponds to the conventional CTC system while computers and terminals of COMTRAC are depicted in the rest of the figure. We have built COMTRAC on the existing CTC system. EDP, PRC and PIC stand for Data Processing System, Route Control System and Passenger Information System respectively. Other two systems to control substations and to control rain/wind gauges are omitted in this figure, though they indeed play important roles in safe and reliable train operation on the Shinkansen. To satisfy the requirements of fault-tolerance, EDP, PRC and PIC have dual, triple and dual computers respectively.

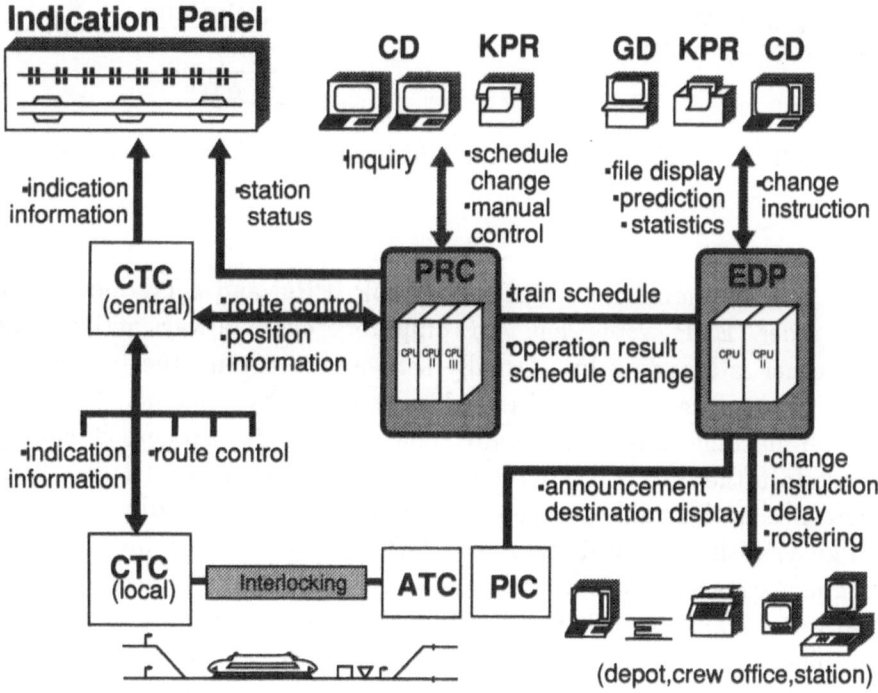

Figure 8: Configuration of COMTRAC Computer Systems

Figure 9 shows the hierarchical and redundant structure of computer

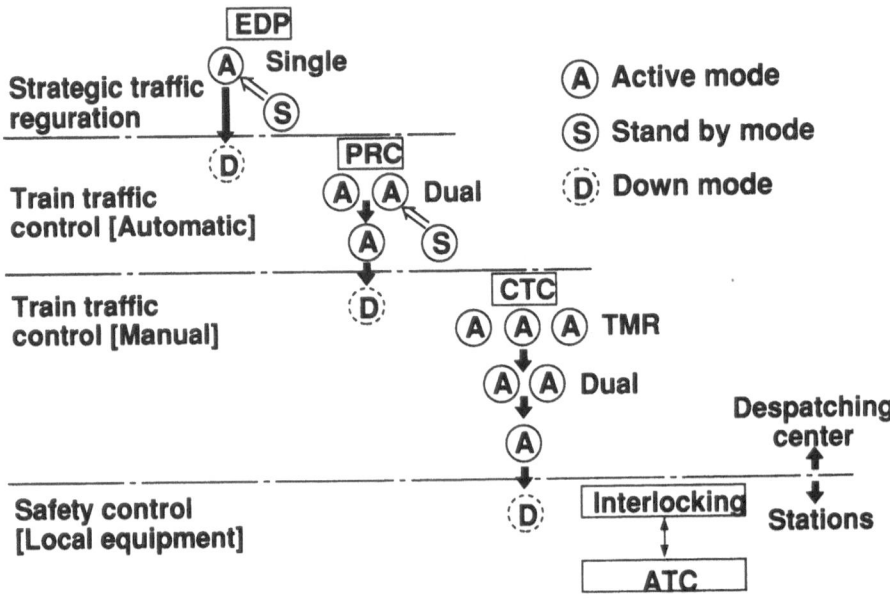

Figure 9: Hierarchical and Redundant Structure of Computer Systems

systems. Availabilities required for EDP, PRC and CTC increase in this order. EDP system has two computers; one is always active while the other is standing by. Usually software to extend the function of the system is developed and tested on the standby machine. When the active machine fails, the standby one takes over. Because it is easy to restore the files in the case of a failure and usually EDP has little influence on the train operation in near future, a cold standby one seems to give satisfactory availability to EDP. PRC has three computers, two of them are active according to the concept "same software on different hardwares" and the other is standby. We can use the standby machine for maintenance and test of the programs as well. The purpose of having two active computers is not for safety but for prevention of a train from running into a wrong track in station. CTC has three computers of TMR (Triple Modular Redundancy). The reason why CTC is required to have the highest availability is that we want to have the possibility left to control the Shinkansen manually using only CTC.

4.3   Functions of COMTRAC System

COMTRAC consists of three levels of functions, schedule generation, traffic regulation and train control. The major functions in each level are described below. They are performed in either EDP or PRC or PIC.

(a) Schedule Generation
Fundamental train schedule is renewed every several years based on the transportation demand. Seasonal or daily train schedules can be compiled from the fundamental schedule by taking seasonal or daily information into account. EDP can almost automatically compile these schedules if the information involved is provided. After train schedule is compiled, EDP can generate car/crew rostering plans. Train schedule and car/crew rostering plan for the day are sent to relevant offices such as stations and car/crew depots. EDP also prepares passenger information for audio/ visual facility in station. Finally EDP transmits the daily train schedule to PRC and passenger information to PIC respectively.

(b) Traffic Regulation
Both EDP and PRC are responsible for traffic regulation. In every ten minutes EDP predicts the state of traffic beyond three hours, while it tries to detect any conflict in car/crew rostering plans. Receiving warnings from EDP, the dispatcher can either deal with the warnings immediately or put them off at his will. All trains in operation are tracked independently by both PRC and CTC. The location of each train is displayed on the indication panel. If PRC detects some disturbance in the traffic and decides to change the train schedule, it inquires the dispatcher to determine finally whether to change it or not. Getting his approval, PRC modifies the train schedule file and sends the information back to EDP. EDP stores temporarily the information in the buffer. Any task which refers to train schedule file deals with the information in the buffer and all the files in EDP relevant to train schedule are modified as well. Final decision for traffic regulation is always left to the dispatcher. The dispatcher can also change the car/crew rostering plan anytime by entering necessary information into EDP. Change information either in train schedule or in car/crew rostering plan will be transmitted immediately to the relevant offices and also to PIC.

(c) Train Control and Passenger Information
PRC always tracks each train by examining the indication data of CTC. When a train approaches to a station or it is the time for another train to depart from the station, PRC sends control data to interlocking device through CTC and offers the inbound/outbound routes for those trains. To give passengers in station the information about the state of traffic, PIC controls the announcement and the arrival/departure signboards in each station. PIC receives regularly once a day the necessary information from EDP. Anytime when the change of train schedule occurs, such information may well be modified to tell the passengers about the state of traffic correctly.

5 Dependable Technologies of the Computer System

In railway safety is the most important requirement. In this paper we would like to include both safety and fault-tolerance in the concept of dependability. We basically realize a dependable computer system by using multiple computers. For fail-safe purpose multiple computers must give identical outputs and for fault-tolerant purpose at least one of them must seldom fail even if all the others should fail.

5.1 The Concept of Fail-Safe in Railway

Fail-safe means that the effect of failure emerges as a safe-side operation. The concepts to realize the safety in railway are as follows;

(a) Explicit definition of both safe-side and risky-side states

(b) Switching output over to safe-side state in case of failure

(c) Redundancy implementation by either hardware or software or both

(d) Using independent modules for redundancy

(e) Immediate detection of data error

(f) Quick identification of the latent faults

The four concepts (c) to (f) seem to be quite common to many fail-safe problems. We believe the first two concepts are quite peculiar to railway and other specific fields like utilization of nuclear reactor. We can define explicitly the safe-side state and the risky-side state for every train control equipment. For instance the red lamp of wayside signal requires the train to stop this side of it. Switch at rest is safe but it is risky in motion. The principle of fail-safe in railway is to stop the train every time it encounters a risk, which is a conclusion of concepts (a) and (b).

5.2 Fail-Safe Technology for Railway[4]

To realize a fail-safe computer system we use multiple computers with identical software. We assume that the probability of the same parts of the multiple computers failing simultaneously is small enough to be neglected. Consequently we can get fail-safe control output from fail-safe comparator.

The concept of FSC (Fail-Safe Comparator) is depicted in Figure 10. Data from the buses are put in the left and right PSRs (parallel/serial register). One "1" and one "0" are added in the last bit positions of left and right PSRs respectively. The contents of PSRs are taken out bit by bit to be sent to the XOR logic by the control unit with a higher clock. There is a two bit shift register. Right shift input of this register is the result of "exclusive or" of the corresponding bits of the two PSRs, while the left shift input is given by the control unit.

At first the left shift input sets the content of this shift register at "01". Because the two PSRs usually have identical bit patterns except the last bit position, the right shift input will remain at "0" until it receives "1" generated as the result of XOR of additional bits. In other words the two shift inputs become "1" only once in each data cycle to produce the rectangle voltage in the output. However if the two PSRs get different data, the right shift input becomes "1" more than once during the same data cycle, leaving "11" in the shift register. In this case a certain dc voltage will be generated on the output terminal, which causes the dc relay to shut down the control output of the computer system.

An example of the application of FSC is shown in Figure 11, which is a fail-safe computer system with bus level data comparison. In this system every pair of data through two buses (input data, control output

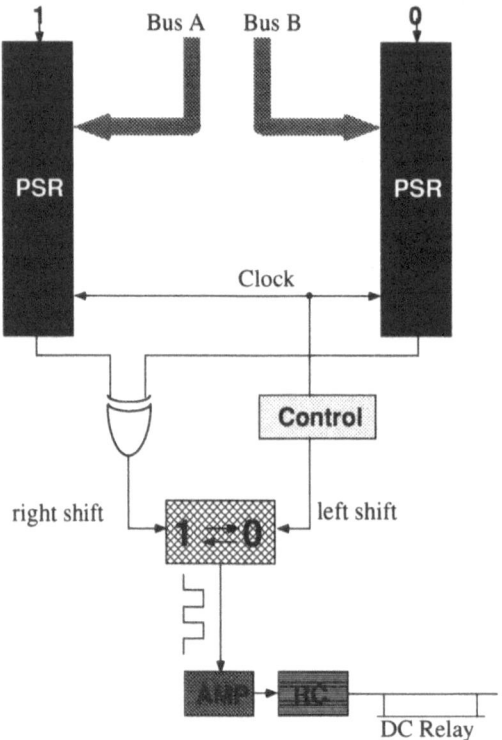

Figure 10: Concept of Fail-safe Comparator

and data from/to memory) are compared by the FSC in every machine cycle. Only when one data coincides with the other, FSC allows control data to go out. Figure 12 shows a block chart of an on-board ATC of TMR structure with fail-safe computer. The two computers in each module are for safety and the three modules are for reliability.

5.3 Highly Fault-Tolerant Technology in PRC

Computers for ATC and interlocking device must be fail-safe. Wirelogic ATC in the Shinkansen has already been replaced in part with computerized ATC. The replacement of wire logic interlocking device with computerized interlocking will also be realized in near future. On the

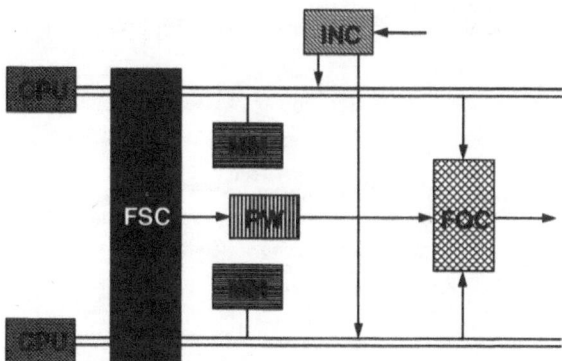

FSC: Fail safe comparator
PW : Power supply
MM: Memory
FOC: Fail safe output circuit
INC: Input circuit

Figure 11: A Fail-Safe Computer System

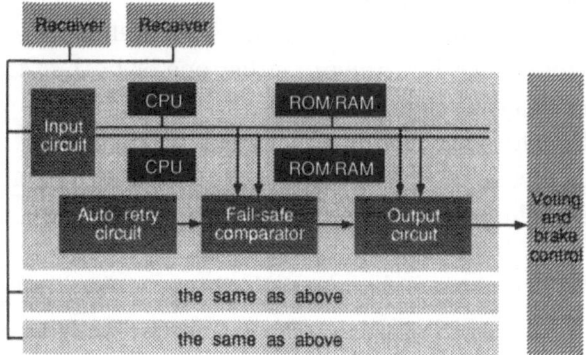

Figure 12: A Computerized ATC

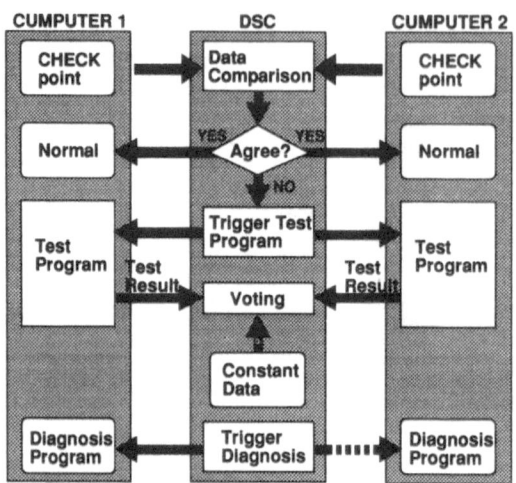

Figure 13: Concept of Dual System Controller

other hand COMTRAC computers need not be fail-safe, because they are indifferent to the safety of the Shinkansen. As we mentioned in section 4.2, the purpose of dual computers in PRC is to avoid wrong route setting of train. For instance, if a stopping train enters tracks without a platform, it will be necessary to shunt the train into the predetermined track. But such shunting operation takes quite a long time and causes delay of all the succeeding trains.

If the control output of one active computer is different from that of the other, something must be wrong in either one or both of the computers. We use DSC (Dual System Controller) to guarantee the proper output and to identify the failed computer. Figure 13 illustrates the function of DSC. DSC is a custom LSI to control the operation of PRC. We define check points in the process of PRC computer, where DSC receives check point data from both of the dual computers and check whether they agree or not. The check points in PRC computer are as follows;

(a) Input

(b) Control Output

(c) Contents of Interrupt

(d) Task Execution Sequence

If two check point data are identical, both computers are normal. Otherwise DSC triggers both computers to execute the test program, which is designed to be a certain combination of the basic instructions of the CPU and yields a predetermined value unless something is wrong with the CPU. DSC has this value as a constant data. Getting the test result about the two computers, DSC decides which of them is in fault by a voting among these three values, i.e., results of computation of both computers and the stored data in DSC.

Concerning the check points the task execution sequence seems to require more explanations. Task synchronization between dual computers of PRC is shown in Figure 14. All tasks in PRC are divided into several task groups depending on the files they refer to. The executability of tasks belonging to the same group is examined periodically in a fixed order. Each task group has a request table which is to be sent finally to DSC. Only the task, of which the results of executability are identical in both computers, will be finally executed. Any task can have access to files of higher task group through a dummy task as shown in the figure.

When both active computers fail simultaneously, it takes some time for the standby one to take over. Meanwhile trains run on and the order of trains may possibly be changed, resulting in a disruption of the train tracking file in PRC. Consequently it is very important to restore this file as soon as possible. To meet this requirement we have another redundancy. The location of train on the indication panel in the General Control Center is maintained independently of PRC by the indication panel controller attached to CTC. The procedure to restore the train tracking file is as follows. First the latest train schedule in EDP is transmitted to PRC. Second the present locations of the trains are sent from the controller to PRC. Then PRC reads the present indication of track circuits and switches from CTC. Finally PRC resumes the train tracking file automatically. Having another controller other than PRC will cost a lot. But it will contribute much to realizing a small MTTR. Without such a controller the dispatcher would have to guess to know the train number for each track occupancy of CTC.

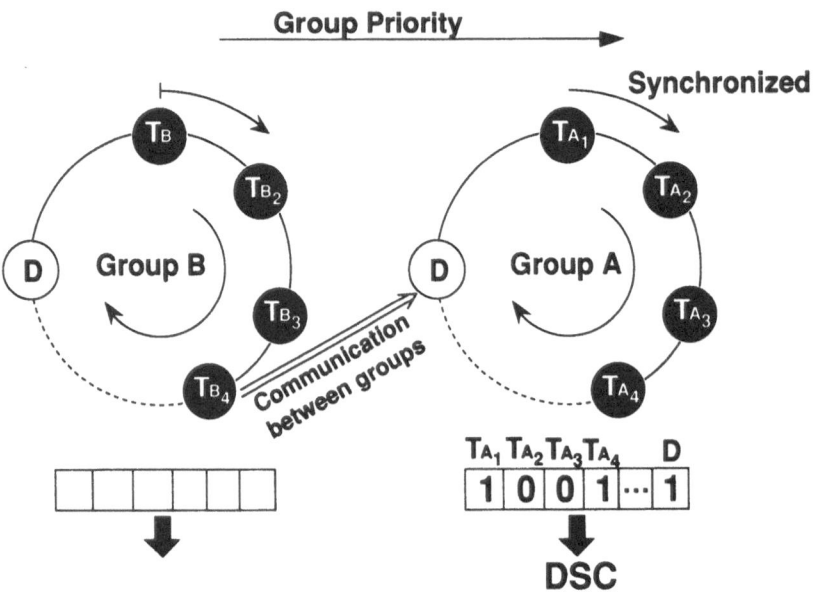

Figure 14: Task Synchronization between Dual Computers

5.4 Robustness and Graceful Degradation

Schedule generation, traffic regulation and train route control/passenger information are the three major functions of COMTRAC. From the fault-tolerant point of view, however, their importance increases in this order. Accordingly we designed the robustness and graceful degradation of COMTRAC, taking the characteristics of these functions into consideration.

EDP generates daily train schedule and daily car/crew rostering plan by batch processing. Consequently PRC and PIC can go on with their job without EDP once they have got the train schedule for the day. When EDP fails, it becomes impossible to transmit the change of passenger information to PIC, and it also becomes impossible to transmit delay information and change of car/crew rostering plan to car/crew depots. Because in most cases trains run exactly on schedule, there are few opportunities of changing the information that has already been transmitted. And PIC in station can control locally the audio/visual facility using original information given in advance.

The traffic may be disturbed by weather, earthquake or equipment failures. In that case we must regulate the traffic in an optimal way. In
COMTRAC both EDP and PRC help the dispatcher in various ways.
The computers may propose a change of train schedule and they can
check the consistency of the tentative instruction to change the train
schedule issued by the dispatcher. For a minor disturbance it is PRC
that proposes a regulation method to the dispatcher. And PRC can
finally change the train schedule by getting his approval and it returns
the new schedule back to EDP. For a major disturbance EDP simulates
the effect of regulation on the traffic beyond several hours. If it finds
some conflict in the car/crew rostering plan, it gives a warning to the
dispatcher and proposes a change of the plan. Although the dispatcher
often accepts the proposal of the computer regarding car rostering, he
seldom admits its proposal regarding crew rostering plan. This is because it is impossible for EDP to draw up fully feasible plans for crew
rostering. The feasible plan must take every condition especially human-
factors into consideration. Anyway PRC is responsible for a rather short
range of schedule, while EDP handles a rather long range planning. Consequently COMTRAC can continue to help the dispatcher regulate the
traffic even when EDP fails. The function that we expect to survive to
the last is train route control.

Although we stated that two active computers are required in PRC to
avoid a wrong route setting, PRC continues to be in service even if the
number of active computers is reduced to one. This is because we could
control train route using CTC terminal but in that case the number of
trains in operation would be reduced to a great extent. Moreover we
implemented a graceful degradation in software to increase the availability. If any task or any module of it fails in the single active computer, it
will be aborted unless it has direct relationship with train route control.
Even if all of the three computers of PRC fail, the dispatcher could set
train route manually using CTC terminal. Each station has duplex local
units of CTC. Therefore if one of them fails, the other will take over.
If both of the local units fail, the station master can set train route by
operating directly the interlocking device.

		function	response time	throughput	reliability
1	analytical model		O	O	
2	train traffic simulator stand alone mode	O			
	cooperation mode	O	O	O	
3	monitor run test	O	O	O	O
4	control run test	O	O	O	O

Table 1: Test Items in System Evaluation and Verification

5.5 System Evaluation and Verification

We have evaluated and verified COMTRAC in all of the following stages;

(a) Analytic Model

(b) Real-time Simulator

(c) Monitor Run Test

(d) Control Run Test

The items to be evaluated or verified are function, response time, through-put and reliability. Test items in each stage are shown in Table 1. Among them TTS (Train Traffic Simulator) was very helpful to evaluate the functions and system performance. Figure 15 shows three ways of system evaluation by TTS.
In the stand alone mode both APS (APplication Software) and TTS were installed in the same computer to test the function of COMTRAC. In the cooperate mode, TTS and APS were installed in different computers to estimate response time and throughput of COMTRAC. Monitor run and control run were the final stages of evaluation and verification of the system. In monitor run PRC receives input directly from CTC and transmits control output to a dummy terminal instead of CTC. In control run both input and control output were real data from/to CTC. In other words PRC performed actually train route control in the Shinkansen.

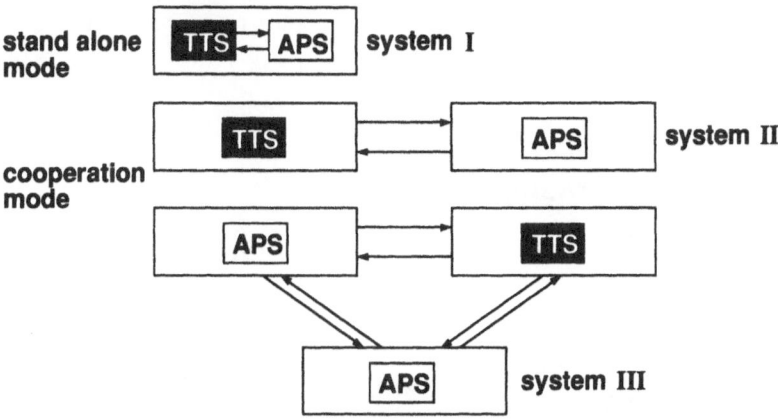

Figure 15: System Evaluation by Train Traffic Simulator

6   Field Data on Reliability and Availability

Figure 16 shows an annual record of faults in EDP. We can find several features from this chart. First of all we notice a decreasing tendency in the number of software faults. On the other hand there is little remarkable change in the number of hardware faults except in 1989. The reduction of faults in software seems to be due to the elimination of the program bug. Recently a system down due to software has never occurred.

Figure 17 shows an annual record of faults in PRC. Both software and hardware seem to have the same tendency as those in EDP. We must pay attention to the fact that the scales of the two charts are different. It seems difficult to remove the remaining bugs completely and a system down due to software may occur occasionally even in the future. But we can claim that no system down of PRC due to hardware proves its high reliability.

Figure 18 shows a change of availability of EDP. Since 1983 it has exceeded the target value of three nine.

Figure 19 shows a change of availability of PRC. It has been improved considerably during the last 15 years and has exceeded the target value of four nine as well. From this chart we may easily guess that in 1970s system down of PRC must be frequent. Indeed it is true that system

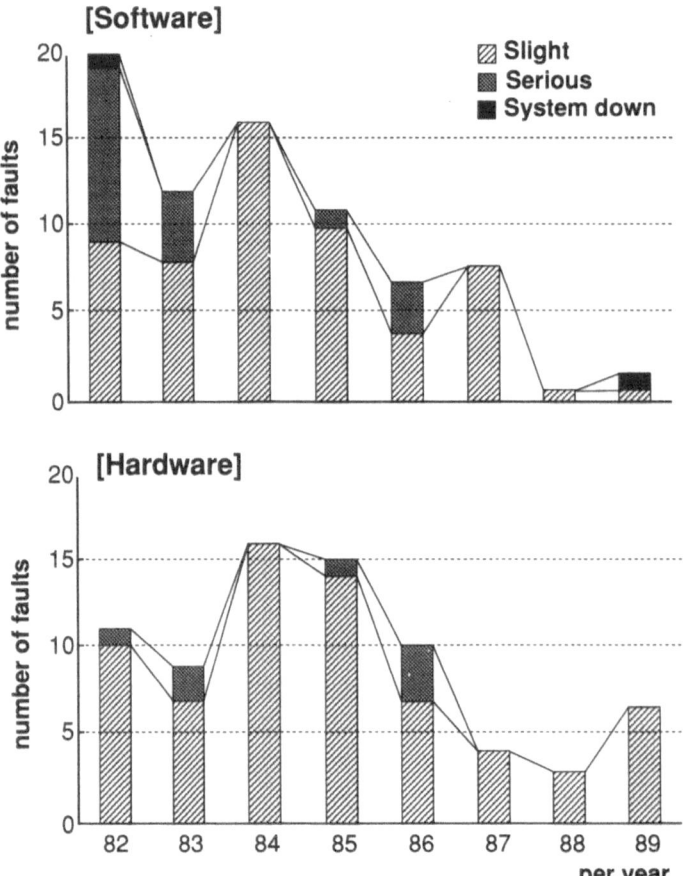

Figure 16: Annual Record of Faults in EDP

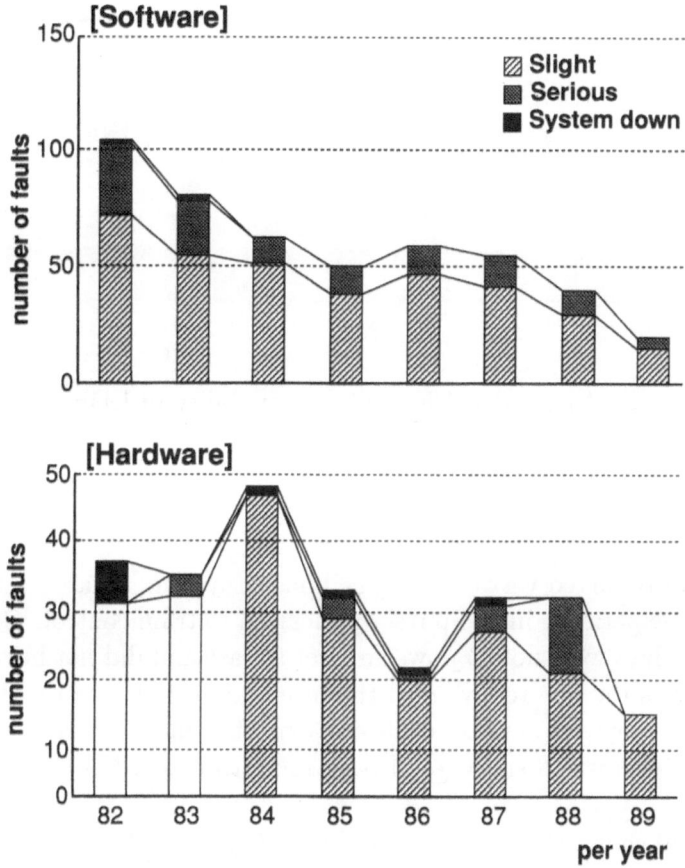

Figure 17: Annual Record of Faults in PRC

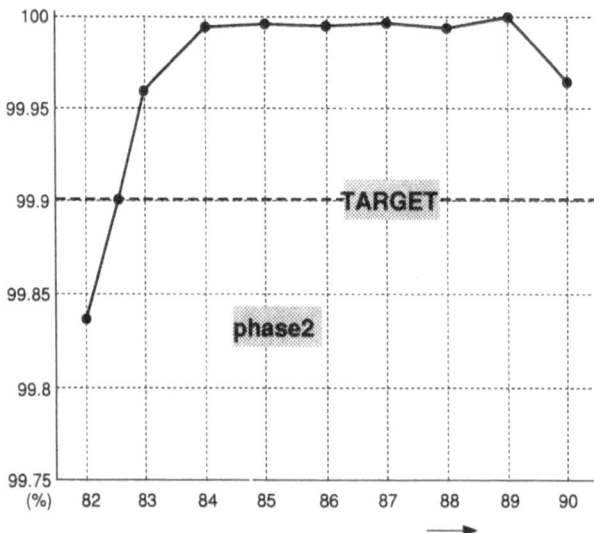

Figure 18: Change of Availability in EDP

down has occurred very often in the phase 1 system. Phase 1 system was
our first experience in computer application to train control. Computers
in those days were not so powerful, not so fast and did not have so large
memory as the one today. And the interactions between the dispatcher
and the computer surpassed all our expectations. The major cause of
system down was a shortage of processing time of CPU.

Figure 20 shows the relative amounts of faults in PRC and EDP in the
phase 2 system on the annual average.

The reliability and availability in PRC for the three phase systems are
summarized in Table 2. Computer hardware today is extremely reliable.
Thus it is the design and production of computer software that will
become still more important in the future. Experience teaches us that
the simpler the software structure can be, the more reliable the software
will be.

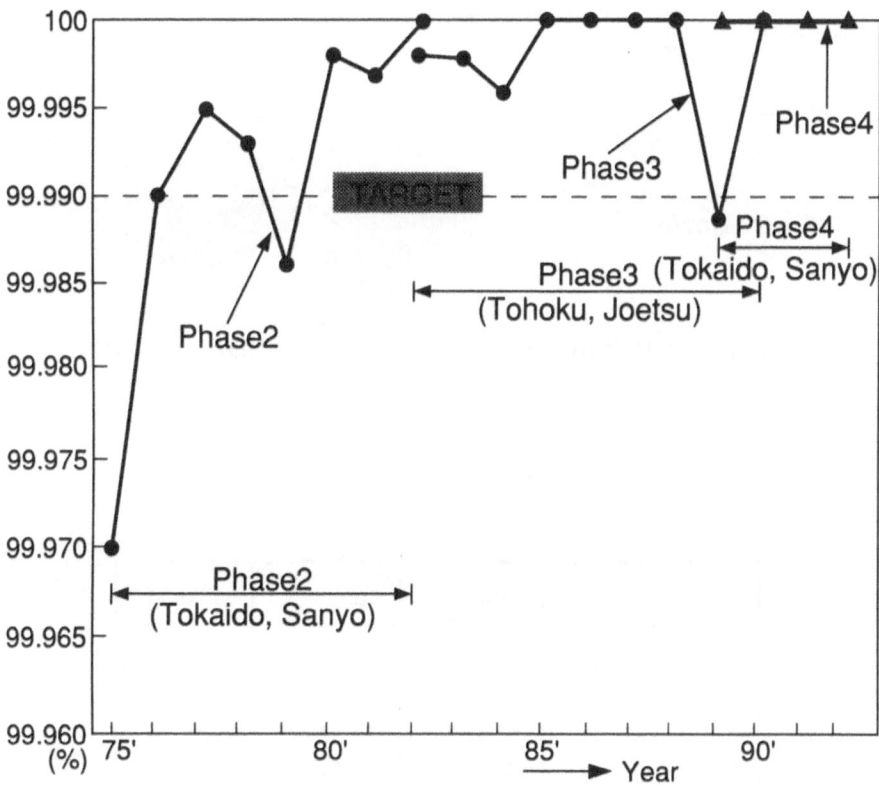

Figure 19: Change of Availability in PRC

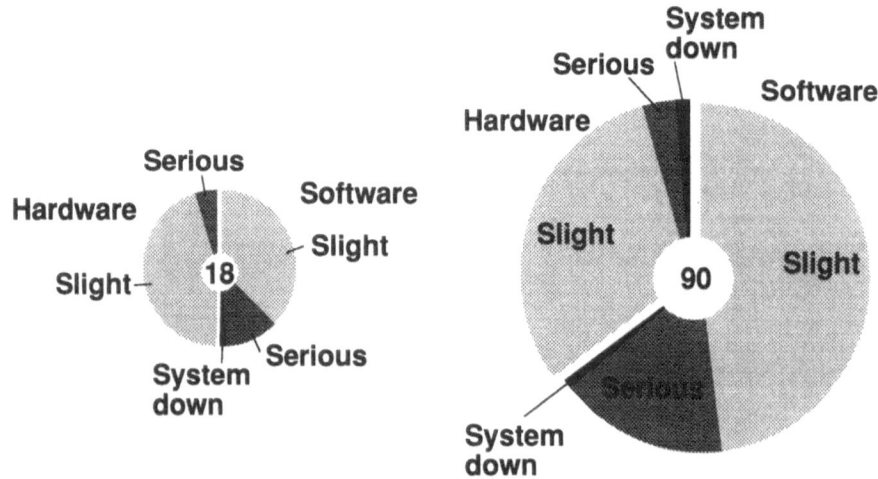

Figure 20: Relative Amounts of Faults in EDP and PRC

	phase2	phase3	phase4
Single System failure [per year]	9.5	1.0	0
MTBF [h]	4,380	14,783	∞
MDT [h]	0.5	0.3	0
Availability [%]	99.991	99.998	100.000

Table 2: Reliability and Availability of PRC

7 Conclusions

Although we can not imagine any reliable train operation in the Shinkansen without COMTRAC, we must point out other aspects of the Shinkansen operation, which also support safety and reliability of the trains. First we must not overlook the maintenance work for track, overhead line, rolling stocks and other equipment of the Shinkansen. Troubles in these facilities may give a lot of damages to the safe and reliable train operation. Second we must mention several warning systems which send information immediately to the dispatcher when some risky phenomenon (big earthquake, strong wind, heavy rain, etc.) happens. These systems together with COMTRAC can help the dispatcher take control of the safe and reliable train operation on the Shinkansen.

Finally we would like to give our view of future computer system in the Shinkansen. COMTRAC today seems over-equipped. It costs a lot and it is time-consuming to add or change its functions. COMTRAC in the future will be a distributed system designed using open system architecture. Especially each station should have its own PRC system. The functions of train route control in future PRC should include route interlocking in station. CTC will be replaced by a standard network. The distributed PRC will be connected to and exchange information through this network. To date replacement of PRC hardware has often accompanied reproduction of PRC software. Thus replacement of the system was very expensive. We expect that the open system technology will allow us to design a system, in which most of the parts will be easily replaced with cheap and simple alternatives.

References

[1] J. R. Group, ed., *High Speed Railways in Japan Present and Future Pamphelet*, 1992.

[2] Y. N.Inada and Y.Hirota, "Shinkansen traffic control and management system - comtrac phase 2," *RAIL INTERNATIONAL*, vol. 14, no. 4, pp. 295–308, 1975.

[3] A. A.Seki and M.Katahira, "New "comtrac" computer-aided traffic control system for the tokaido-sanyo shinkansen," *Proc. of 2nd Int.*

Conf. on Computer Aided Design, Manufacture and Operation in the Railway and other Advanced Mass Transit Systems, pp. 231–242, 1990.

[4] H. K.Akita, T.Watanabe and I.Okumura, "Computerized interlocking system for railway signaling control:smile," *IEEE Transactions on Industry Applications*, vol. 1A-32, no. 4, pp. 826–834, 1985.

Real-Time Systems

Exception Handling in Real-Time Software from Specification to Design

R. D. LEMOS, A. SAEED, A. WATERWORTH

Department of Computing Science,
University of Newcastle upon Tyne, NE1 7RU, UK

Abstract

In this paper we present a systematic method to cope with exception handling in the specification and design of real-time software. Guidelines for the construction of programs, directly, from a formal specification are also provided. The method proposed is an object-based approach which makes use of Petri Nets with Objects, and an object-based design notation which incorporates facilities for exception handling and active objects. To illustrate the proposed approach a case study based on a train set is presented.

Keywords: real-time systems, exception handling, formal models, time modelling, object-based systems.

1 Introduction

The systematic method for the specification and design of real-time software discussed in this paper is presented within the context of a framework for the requirements analysis of safety-critical systems [1]. (Here, we consider a specification to be the product of the requirements analysis.) The basic aim behind the framework is to subdivide the whole problem into smaller domains where the analysis of the requirements can be simplified, thereby leading to more accurate requirements specifications.

In this paper, we move the framework on to the next stage in software development by describing a link from the specification to design of real-time software. We are not concerned with transformation techniques that would allow the designer to derive automatically the code of a program from a specification. Instead, we provide adequate notations to deal with exception handling in the specification and design, and some guidelines which aid in the construction of a program from a formal specification. The formalism to be used for the specification is Petri nets with Objects, an high-level Petri net, with an extension to support the representation of timing constraints. The utilization of a formal notation in the requirements phase allows us to perform formal verification of the specifications, before we proceed to the design phase, for which we suggest an object-based design notation which incorporates facilities for exception handling and active objects.

The approach followed in this paper is similar to the work described in Cristian [2] and Bidoit et al. [3]. The approach described by Cristian is based on abstract data types; operations on the data are specified as state transition relations, in which the relation is defined by the precondition and postcondition of the operation, for the design a Pascal-like programming language with a class construct like SIMULA is used. In the approach described by Bidoit et al, an algebraic specification language (PLUSS) is employed for the specification of input or output data types, and for the construction of programs, expressed in Ada, a set of decomposition schemes are provided.

The approach to be followed in the development of software for safety critical-systems is based on maintaining a clear separation of the mission from the safety requirements. Whether it is feasible to separate the mission from the safety requirements for a particular safety-critical system depends on the ease with which a distinction can be drawn between safe and unsafe states.

Below we define those specifications produced by applying the framework for requirements analysis, which are most relevant to the work presented in this paper:

safety constraint - is a condition imposed on the system that is the negation of a hazard modified to incorporate safety margins;

safety strategy - is a way of maintaining a safety constraint, and is defined as the conjunction of a set of conditions over the physical process;

robust safety strategy - is a refinement of a safety strategy that takes into account the limitations of sensors and actuators;

safety controller strategy - is a refinement of a robust safety strategy incorporating some of the controller components and their relationship with other components, such as sensors and actuators.

The general structure we adopt for safety-critical systems consists of three different components: the *operator*, the *controller*, and the *physical process* (or *plant*). The controller has two interfaces, the *plant interface* and the *operator interface*; the plant interface is made up of sensors and actuators. The system has an *environment* which is part of the rest of the world which may affect, or be affected by, the system. To reflect the separation of the safety from the mission, the operator and controller are partitioned into the *safety operator* and *mission operator* and *safety controller* and *mission controller*. The safety controller and the safety operator are the components which must ensure that the system does not enter into a state which can lead to an accident.

The work presented in this paper is restricted to those failures in sensors and actuators which might affect the safety of the system. Other failures which could be treated using exception handling, such as violations of assumptions over the behaviour of the physical process, are not considered. When either a sensor or an actuator failure occurs, the safety controller should take corrective measures that ensure the safety constraints are maintained.

The rest of the paper is organised as follows. In the next section a model for the behaviour of real-time systems is presented. In section 3, we discuss how to model the behaviour of sensors and actuators in terms of their standard, exceptional and failure behaviour. Sections 4 and 5 describe, respectively, the notations to be employed during specification and design. Also in section 5 we describe how the development process of the two phases should be combined. In section 6, a case study based on a train set is discussed. Finally, section 7 contributes some concluding remarks.

2 Modelling Real-Time Behaviour

When modelling the behaviour of real-time systems we are concerned with events and actions where the time reference is explicit, this necessitates the utilisation of an underlying time structure to model the flow of time.

2.1   Time Structure

For the formal definition of time we adopt the concept of point structures that can be specialized to either a dense or a discrete time structure. On one hand, dense time structures are more convenient to express the physical laws that describe the behaviour of the real world, and on the other hand, discrete time structures seem a more convenient and expressive model of time for discrete computation. A dense time structure is employed during the early stages of the requirements phase, while a discrete time structure is used in the later stages of the requirements phase and the design phase.

A general point time structure $\mathcal{T} = <\mathbf{T}, <, \Delta, - >$, that is a strict partial order, linear and for which adjacent points are separated by Δ ($\in \mathbf{R}_+$) has been defined elsewhere [4]. For such a time structure the notion of a *timeline* is defined as a sequence of points connected by an arc, and an *interval* is simply a subset of $\mathbf{T}$ that represents an uninterrupted stretch of time. The *dense time structure* used here is defined as a special case of $\mathcal{T}$: $\mathcal{T}_{DE} = lim_{\Delta \to 0} <\mathbf{T}, <, \Delta, - >$ which is isomorphic to the reals. The *discrete time structure* used here is defined as: $\mathcal{T}_{DI} = <\mathbf{T}, <, 1, - >$ which is isomorphic to the integers.

2.2   Event/Action Model

For the analysis of the timeliness requirements we employ an *event/action model* (E/A model) based on the work of Jahanian & Mok [5], with some features that make it more applicable to requirements analysis. Namely, the model is flexible enough to support both discrete and dense time structures [4], has the potential of obtaining both descriptive and operational semantics, and has the ability to depict graphically the timing analysis.

The model consists of two basic notions, events and actions, and two Boolean functions that represent time uncertainties related to temporal behaviour of events and actions. An *event* is a temporal marker of no duration, represented as a cut in the timeline. An *action* is a basic unit of activity manifested by a *start event* marking the initiation of an action and a *finish event* marking the completion of an action. We introduce the concept of an "event-occurrence function" $E(t, i)$ that returns the value one if and only if the ith occurrence of the event occurs at time point t. Similarly, we introduce an "action-execution function" $A(t, i)$ which returns one if and only if time point t is contained within the ith execution of the action.

The specification of time uncertainties are represented by means of a "critical utility function" $U(t, i)$ that can return a maximum (1) or minimum (0) value. The utility function for an action is defined in terms of four time attributes, earliest start time - t_{est}, latest start time - t_{lst}, earliest finish time - t_{eft}, and latest finish time - t_{lft}. For an event, $U(t, i)$ returns one if and only if the ith occurrence of the event is contained within the interval $[t^i_{est}, t^i_{lft}]$. The time uncertainties for an action are defined by the functions: $U_{\uparrow A}(t_{sta}, i)$ which returns one if and only if, for the ith execution, the start event occurs in $[t^i_{est}, t^i_{lst}]$, and $U_{\downarrow A}(t_{fin}, i)$ which returns one if and only if the finish event occurs in $[t^i_{eft}, t^i_{lft}]$.

3 Behaviour of Sensors and Actuators

In this section we are concerned with the analysis of sensor or actuator failures that have an adverse affect on the safety of the system. This involves specifying the anticipated values (**AV**) and unanticipated values (**UV**) of physical variables, in terms of their range; and the standard, exceptional, and failure behaviour of the sensors and actuators. Here we sketch a model for the behaviour of such a component. The model follows an approach adopted for specifying the standard and exceptional behaviour of software components [6].

For component CM let **D** ($=$**AV** $\cup$ **UV**) be the set of values that may be inputs for CM and **R** the set of values returned by CM. The behaviour of CM is defined in terms of partial relations over **D** $\times$ **R**. The standard behaviour is defined by relation $SS(CM)$:

$SS(CM) = \{ (d, r) | d \in$ **AV** $\wedge r$ is an intended outcome for $d \}$.

For component CM we specify the set of exceptions as **E**. For each exception $e \in$ **E** we specify the exceptional behaviour for CM by the relation $ES_e(CM)$:

$ES_e(CM) = \{ (d, r) | d \in$ **AV** $\wedge r$ is an intended outcome for d for $e \}$.

We will say that component CM has failed if its behaviour is not con-

tained within the relation $SS(CM)$ or an relation $ES_e(CM)$. The failure behaviour is specified by the relation $FS(CM)$:
$$FS(CM) = \{ (d,r)|d \in \mathbf{AV} \wedge (d,r) \notin SS(CM) \wedge$$
$$\forall e \in \mathbf{E}: (d,r) \notin ES_e(CM) \}.$$
The failure behaviour of a component is identified by conducting a failure modes and effect analysis (FMEA). A FMEA is used to analyse all component failure modes and to identify the resulting affect on the system [7]. We denote the set of anticipated failure modes as $\mathbf{FM}$; and for each failure mode $af \in \mathbf{FM}$, we specify the failure behaviour by the relation $FS_{af}(CM)$:
$$FS_{af}(S) = \{ (d,r)|d \in \mathbf{AV} \wedge r \text{ is an outcome for } d \text{ in } af \}.$$
The anticipated failure behaviour will be a subset of the failure behaviour:

$$\bigcup_{af \in \mathbf{FM}} FS_{af}(CM) \subseteq FS(CM).$$

In the above specifications, we have concentrated on relationships in the value domain. However, for many practical systems it will be necessary to extend the specification to incorporate the relationships that exist in the time domain [8].

4   The Specification Notation

During the requirements phase we adopt two formal notations, Timed History Logic (THL) [9] and Petri nets with Objects (PNO) [10]. THL is based on a dense time structure $(\mathcal{T}_{DE})$, and consists of three main concepts: *histories*, *relations* and *modes*. A history of a system represents a possible behaviour (i.e. a sequence of values taken by the state variables) of the system during its lifetime. The universal history set is the set of all such histories. *Invariant relations* are predicates used to express relationships over the state variables which hold at every time point in the lifetime. *History relations* are predicates used to express relationships which hold during every interval included within the lifetime. A specification can be expressed as a set of invariant and history relations, these specify the subset of histories of the universal set that are acceptable.

Petri nets are mainly used for the modelling and analysis of discrete-event systems which are concurrent, asynchronous and non-deterministic. The use of PNO adds to the modelling power of Petri nets the formal treatment of *individuals* (i.e. the notion of token identity) and their changing properties and relations [11]. The incorporation of time in PNO is based on a discrete time structure ($\mathcal{T}_{DI}$) and will follow what has been proposed for ER nets [12]. To each tuple we associate a timestamp which represents the time point at which the tuple was produced, this depends on the relational expressions associated to the transitions. The E/A model may be represented in terms of PNO in a form similar to that adopted for another type of high-level Petri net [4].

For the specification of the safety controller we adopted an object-based notation which is based on the following notions. A class of objects is identified by its name and has a structure which includes a data structure (or attributes) and a set of operations. Each attribute has a name and a type. The value of an object is processed through the operations of the object, called its methods. The attributes and operations of an object can either be declared as public or private. Those methods that implement the robust safety strategy are specified in terms of THL formulae. The sequence in which operations are performed by an object is specified as an PNO. Here we do not show the verification of the PNO implementation of the robust safety strategy against the THL formulae, the verification could be performed by following the approach suggested in [13].

5 The Design Notation

The design notation adopted is based upon that used in [14]. This is an object-based notation that offers support for *active objects* (i.e. objects containing one or more autonomous internal threads of control) and incorporates a "traditional" exception handling mechanism similar to that used in [6]. A *single-level termination* model of exception handling is assumed.

Object methods are assumed to be executed as atomic transactions, however we will only be relying upon the support for serializability (or, more specifically, non-interference of concurrent operations) which the transaction mechanism provides, since error recovery will be provided solely by means of the exception handling mechanism.

Below we show a simple object declaration in our notation. It is divided into three basic parts. First of all declarations for the private, internal state variables of the object are given, followed by the declaration of the object's private and public methods (in any order). Finally, each object may contain one or more processes, which either carry out simple background processing or dictate the order in which methods are executed.

```
OBJECT Example IS

ATTRIBUTE
        // State variables.

METHOD P(...) SIGNALS e1, e2
begin
        ...
        do
                Some_Object.Q(...) [ Fail :       signal e1 ] ;
        within (EST,LST):(EFT,LFT)
                                    [ Early_Start : signal e2 ;
                                            Late_Start : ... ] ;
                                    [ Early_Finish : ... ;
                                            Late_Finish : ... ] ;
        ...
end [ signal Fail ] ;            // Default exception handler.
        ...
        // Other methods.
        ...
PROCESS                                          // Internal process.
        cycle
                ...
                ...
        end cycle ;

END.
```

A first implementation of a compiler for the design notation would probably be in the form of a pre-processor for an existing language such as Ada. This would require some work to be carried out to provide appropriate run-time support in some areas, for example, providing a mechanism for default exception handlers. However, in principle this should be feasible and such a mapping would then introduce the potential for validation (by means of symbolic execution) of the programs which are produced [15].

Mapping Specification to Design

The mapping between the specification of the system, in terms of class skeletons with a sequence of operations modelled in PNO, and the design of the system can be achieved as follows. First of all, places in the net correspond to *dynamic* state variables. That is, state variables which are subject to change during the course of execution of the software. Incoming edges to a place represent the writing of a value (or values) into the state variable corresponding to that place, while outgoing edges represent the destructive reading of a value (or values) from the appropriate state variable. Labels on the edges give the abstract semantic types of the values which are to be passed, from which the concrete type of a given value can be derived by reference to the class skeleton.

The preconditions placed upon transitions in the net correspond directly to boolean conditions in the program which must be satisfied before an action can proceed. Preconditions marked with a double-edged line represent the "enabling condition" for the raising of an exception, while actions marked in such a way give the handler for that exception. The evaluation of conditions in the program typically takes the form of reading a value from one state variable and performing a search for a matching value in some other state variable. A new state variable value may also be produced if a successful match is found. This pattern of activity corresponds directly to the relationship between places, preconditions and transitions in the PNO itself.

Finally, note that *static* state variables (i.e. those that are assigned values during program initialization and that do not change during program execution) do not correspond to places in the net. However, their names and types can be obtained from the labelling on the PNO and the types defined in the class skeleton.

6   Case Study: Train Set

With the aim to illustrate the concepts introduced so far, a train set
was selected as a case study. Within this case study the train set cross-
ings described below raises safety-critical issues that are similar to those
found at the traditional level crossing (i.e. road-rail). In this section we
exemplify a systematic way how to handle exceptions during specification and design of software.

Conceptual Analysis

The physical process consists of two track circuits Cp and Cs, and two
types of trains - primary (Tp) and secondary (Ts). The circuits are
divided into sections and there are two separate crossing sections at
which they intersect. Trains of type Tp travel around circuit Cp and
trains of type Ts travel around circuit Cs; both types of train travel
in one direction (clockwise) only. The primary trains must always take
priority over the secondary trains at the crossing sections. Specifically,
a primary train must not be made to wait for a secondary train at
a crossing section. Each circuit has thirteen sections, with two trains
running on each circuit; the longest train is shorter than the smallest
section. The crossing sections are sufficiently far apart, to allow the
safety analysis of one crossing section to be unaffected by the presence
of the other. The circuits Cp and Cs, and the crossing sections are
illustrated in Figure 1.

General Model

The type of circuit is denoted by $c \in L, L = \{p, s\}$, the crossing section by $r \in R$, R $= \{a, b\}$, the trains which run on Cc are denoted
by $x, y \in Tc = \{1, ..., Ntc\}$ and the sections of Cc are denoted by
$i, j, k, m, n \in Sc, Sc = \{0, ..., Nsc\}$. Addition $\oplus$ and subtraction $\ominus$ on
circuit section numbers are performed modulo the number of sections
of the circuit. The danger zone on circuit Cc for $CC(c, r)$ is defined
as: $DZ(c, r) = \{CC(c, r), CC(c, r) \oplus 1 \}$. The danger zones $DZ(p, r)$
and $DZ(s, r)$ for a crossing section $CC(c, r)$ are illustrated in figure 1.
The behaviour of the physical process, sensors and actuators and the
recorded position of a train are specified in THL using the state variables below.

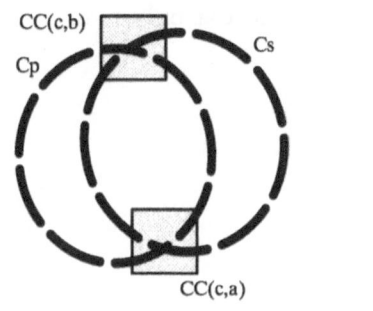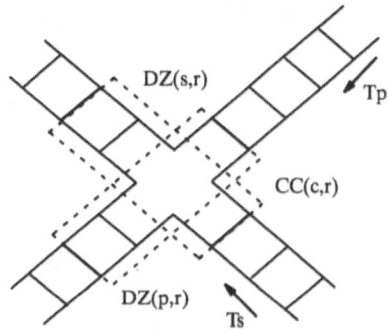

Figure 1: The train set circuits and the crossing section

No.	Name	Range	Comments
p1	*Ptrain*	$Sp^{Ntp} \times Ss^{Nts}$	The position of each train expressed as a section, nuber that is, the section containing the front of a train.
p2	*Rtrain*	$Rp^{Ntp} \times Rs^{Nts}$	The reservation sets of the trains, where $Rc = P(Sc)$.
p3	*Sens*	$\mathbf{B}^{Nsp+1} \times \mathbf{B}^{Nss+1}$	The state of each sensor expressed as a truth value.
p4	*Act*	$\mathbf{B}^{Nsp+1} \times \mathbf{B}^{Nss+1}$	The state of each actuator expressed as a truth value.
p5	*Pos*	$Sp^{Ntp} \times Ss^{Nts}$	The position of each train as observed by the sensors.
p6	*Shut_Down*	$\mathbf{B}$	The operatioal status of the train set, i.e. Shut_Down holds when all trains must be stopped.

For example, $Ptrain(c, x)$ denotes the state variable for the position of train x on circuit Cc. State variables are taken to be functions over time to their respective ranges.

Safety Specifications

Here we present the accidents, hazards, safety constraints and safety strategies associated with the train set example. The safety constraints and the safety strategies are stated informally; the formal specifications and the respective analysis were presented elsewhere [13]. From the various accidents possible on the train set, we consider only two.

Accidents

AC_1 - trains of the same type collide;

AC_2 - trains of different type collide.

Hazards

$HZ_{1,1}$ - some part of any two trains are in the same section.

$HZ_{2,1}$ - some part of a primary train and a secondary train are in the same crossing section.

Safety Constraints

$SC_{1,1}$ - for any two trains there must be at least one section between the sections containing the fronts of the trains.

$SC_{2,1}$ - either the front of no primary train is in a danger zone $DZ(p,r)$ or the front of no secondary train is in a danger zone $DZ(s,r)$.

Safety Strategies

$SS_{1,1,1}$ - the basic rules for the strategy for $SC_{1,1}$ are:

  ssa. for any train, the current section (i.e. the position of the front of the train) and the section behind the current section must always be reserved;

  ssb. no section can be reserved by more than one train.

$SS_{2,1,1}$ - the basic rules for the strategy for $SC_{2,1}$ are:

  ssa. if any train x on circuit Cc is in a danger zone then the crossing section contained within that danger zone is reserved (on circuit Cc);

  ssb. section $CC(p,r)$ and section $CC(s,r)$ cannot both be reserved;

ssc. if $CC(s, r)$ is not reserved by a secondary train x then it cannot become reserved for train x while it is not in section $CC(s, r) \ominus 1$ or the maximum duration for all parts of train x to approach and pass danger zone $DZ(s, r)$ from section $CC(s, r) \ominus 1$ (when it is moving) is at least the minimum duration for any primary train y to approach $DZ(p, r)$.

Sensors and Actuators

At the beginning of every section there is a sensor that detects the presence of a train, and for each train there is an actuator that can stop the train within any section. A sensor may fail and not detect the presence of a train - "miss a train", or detect the presence of a train that does not exist - "ghost train". An actuator may either fail never stopping a train when required or stopping a train inadvertently.

In this example we are only concerned with those sensor and actuator failures which can affect the safety, such as the sensor failure that misses a train and the actuator failure that fails to stop the train when required. In the following we give the standard specification and failure specification for failure mode "miss a train" (mt) of the sensor (S) for section i of circuit c with domain Sc^{Ntc} and range **B**.

SS - sensor S for section i on circuit Cc detects a train x if and only if the front of a train is in the section.

$$SS(S) = \{(Ptrain(c, x), Sens(c, i)) |$$
$$Sens(c, i) \Leftrightarrow \exists x \in T_c : Ptrain(c, x) = i\}.$$

FS_{mt} - sensor S for section i on circuit Cc fails to detect a train when the front of a train is in the section.

$$FS_{mt}(S) = \{(Ptrain(c, x), Sens(c, i)) |$$
$$\neg Sens(c, i) \wedge \exists x \in T_c : Ptrain(c, x) = i\}.$$

For sensor failures we assume an upper bound on the maximum number of consecutive sensor failures $(mcsf = 2)$, and for actuator failures we assume that once a failure occurs, the failure is permanent and the controller ceases to have control over the train.

In the following we give the standard specification and failure specification for failure mode "never stops a train" (ns) of actuator (A) for train x on circuit Cc with domain $\mathbf{B}$ and range Sc^{Ntc}.

SS - while the actuator A for train x on circuit Cc is set the train cannot enter a new section.

$$SS(A) = \{(Act(c, x), Ptrain(c, x)) | \forall t \in [T_0, T_1] : Act(c, x)(t) \Rightarrow$$

$$\forall t \in [T_0, T_1] : Ptrain(c, x)(t) = Ptrain(c, x)(T_0)\}.$$

FS_{ns} - train x on circuit Cc moves into a new section while A is set for the train.

$$FS_{ns}(A) = \{(Act(c, x), Ptrain(c, x)) | \forall t \in [T_0, T_1] : Act(c, x)(t) \Rightarrow$$

$$\exists t \in [T_0, T_1] : Ptrain(c, x)(t) \neq Ptrain(c, x)(T_0)\}.$$

To obtain a robust safety strategy we must perform first the analysis of a safety strategy in terms of sensor and actuator failures. The safety strategies $SS_{1,1,1}$ and $SS_{2,1,1}$ must be modified to compensate for sensor failures by increasing the number of sections that must be reserved for a train. This means that the rule ssa of safety strategy $SS_{1,1,1}$ must be modified to ensure that the $mcsf \oplus 1$ sections immediately behind the current section of a train are also reserved; whenever the upper bound on the number of consecutive sensor failures is greater than $mcsf$ the whole system is shut down. In terms of rule ssa of safety strategy $SS_{2,1,1}$ the danger zones must be modified to include the $mcsf$ sections that immediately precede the crossing section. As far as actuator failures are concerned, once an actuator failure is localized, that failure is considered to be a permanent failure and as such, the whole system is shut down. In the following for simplicity we present only the specification of the robust safety strategy for the circuits.

$RSS_{1,1,1}$ - the basic rules for the robust safety strategy for $SS_{1,1,1}$ are:

    rssa. for any train, the current section as observed by the sensor and the $mcsf \oplus 1$ sections behind this section must always be reserved;

$\forall c \in L : \forall x \in T_c :$
$\{Pos(c,x) \ominus (mcsf \oplus 1), ..., Pos(c,x)\} \subseteq Rtrain(c,x);$

rssb. no section can be reserved by more than one train;

$\forall c \in L : \forall x, y \in T_c : Rtrain(c,x) \cap Rtrain(c,y) = \emptyset;$

rssc. if the number of consecutive sensor failures is greater than *mcsf* when the observed position of a train is updated then the system must be *Shut_Down*;

$\forall c \in L : \forall x \in T_c : Pos(c,x)(T_0) \oplus mcsf < Pos(c,x)(T_1) \land$
$\forall t \in [T_0, T_1] : \forall i \in \{Pos(c,x)(T_0), ..., Pos(c,x)(T_1) \ominus 1\} :$
$$\neg Sens(c,i)(t) \Rightarrow Shut\_Down(T_1).$$

rssd. if the recorded position of a stopped train is updated the system must be *Shut_Down*;

$\forall c \in L : \forall x \in T_c : \forall t \in [T_0, T_1] :$
$Act(c,x)(t) \land Pos(c,x)(T_0) \neq Pos(c,x)(T_1) \Rightarrow Shut\_Down(T_1).$

6.1 Controller Specification

In this section we are concerned with the specification of the safety controller strategies for the case study of the train set. The approach followed in modelling the system is to maintain a clear separation between models of the physical process and the safety controller, even though they must cooperate whenever an action is performed [16]. The advantages of adopting this approach are that both models can be independently developed and modified, and the state of the physical process can be seen to correspond to the sequence of control commands issued by the safety controller.

Model of the Physical Process

The development of an accurate model of the physical process is essential to the specification of the safety controller strategies in the sense that, one of the techniques employed in the validation of the specifications is based on the animation of the models of the physical process and the safety controller [1]. In this paper, because we are only concerned with those sensor and actuator failures that can affect safety, the model of the physical process is restricted to the above mentioned failure situations of the sensors and actuators. Also, to limit complexity, we model only a single circuit. This simpler model can be extended to the complete case study.

The PNO model of a circuit of the train set which takes into account failures in sensors and actuators is shown in figure 3. The places of the PNO model of a train set circuit are the following:

$ICP < tn, snj >$ - train tn is allowed to enter section snj;

$IPC < snj >$ - a train has entered section snj;

$S < tn, snj/sni, sf, tstp(Ssnj) >$ - train tn is in section snj or sni, and sf consecutive sensor failures have occurred, $tstp(Ssnj)$ is the timestamp when the train entered the section snj;

$SAF < tn, snj, af, tstp(Ssnj) >$ - train tn is in section snj, and an actuator failure af might occur;

$WTC < tn, snj, tstp(Ssnj) >$ - train tn is in section snj, and has permission to enter into the next section.

The time relation $transSsnj$ associated with transitions $t1$, $t2$ and $t3$ has the purpose to model the time interval that a train takes to pass a section. For that, we associated a minimum and a maximum velocity to the trains these impose the maximum and minimum time, respectively, for a train to pass through a section, as shown in figure 2. The utility function $ISsnj$ represents a train travelling across section snj. The time relation $transWTC$ associated with transitions $t4$ and $t5$, represents the time interval during which the train has to "start to stop" to ensure it can stop within a section. The transitions $t4$ and $t5$ should not fire during the time interval $[t_{est}|_{ST1}, t_{lft}|_{ST1}]$, as shown in figure 3, by $ST1$ the utility function of the train starting to stop in section 1. (The time attributes associated with $ISsnj$ could be used as labels for exception handlers when the train does not meet the timing constraints of the utility function.)

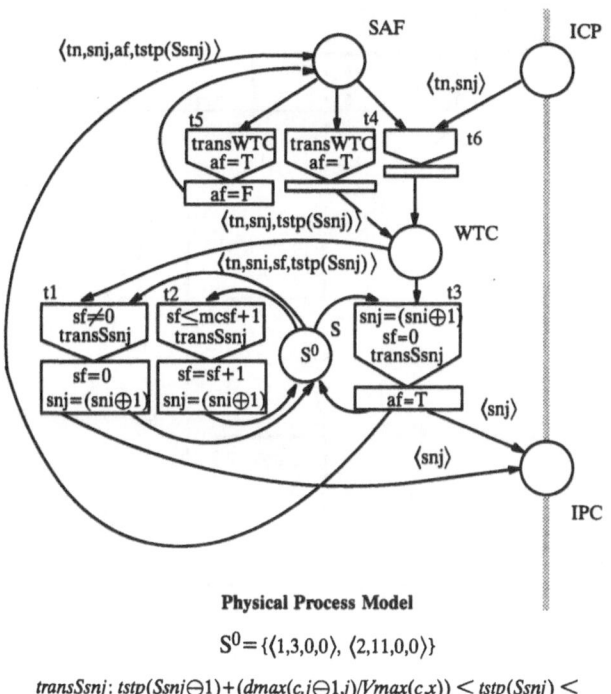

Physical Process Model

$$S^0 = \{\langle 1,3,0,0 \rangle, \langle 2,11,0,0 \rangle\}$$

$transSsnj$: $tstp(Ssnj \ominus 1) + (dmax(c,j \ominus 1,j)/Vmax(c,x)) \leq tstp(Ssnj) \leq$
$tstp(Ssnj \ominus 1) + (dmax(c,j \ominus 1,j)/Vmin(c,x))$, for $j=0$ to 12

$transWTC$: $tstp(WTC) > (tstp(Ssnj) + t_{lft} \mid _{STsnj})$

Figure 2: The PNO model of the train set circuit with sensor and actuator failures

Model of the Safety Controller

The model of the safety controller, as already mentioned, is based on the refinement of a robust safety strategy incorporating some of the controller components and their relationship with other components.

Based on the above robust safety strategies, the safety controller strategy is captured by the following object classes: **Section**, **Circuit**, **Danger_Zone**, and **Train**, as indicated in figure 4. For each section of a circuit there is an instance of the class **Section**, the same applies to the objects of classes **Train** and **Danger_Zone**. The safety controller strategies that avoid hazards occurring at the circuit level and at the crossing

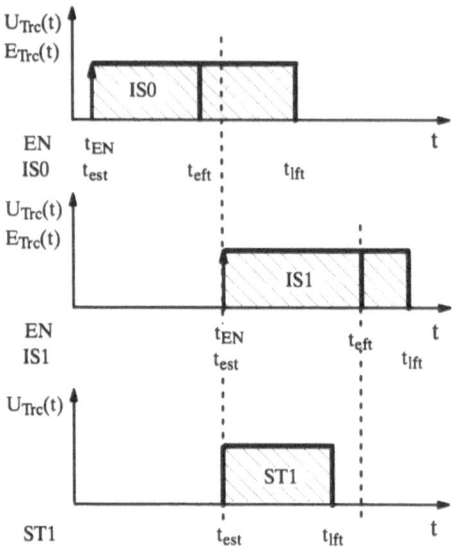

Figure 3: E/A model of a train on the circuit

section level are incorporated in the classes `Circuit` and `Danger_Zone`, respectively. Also, within these two classes we provide the mechanisms which handle the exceptions that are raised due to sensor and actuator failures. For the specification of the classes we adopt the structure already introduced. In the following we proceed to explain in more detail the specification of these classes, excluding the `Danger_Zone` due to space limitations.

The object instances of class `Section` detect when a train enters a section, and inform the `Train_in_Section` methods of the objects `Circuit` (`C[mc].Train_in_Section(ms)`) and `Danger_Zone(DZ[a,b].Train_in_Section(mc,ms,D))`. The specification for class `Section` is presented below.

```
CLASS Section

TYPE
Type_of_Circuit: {p, s} ;
        Circuit_Addresses: array [p, s] of Circuit ;
```

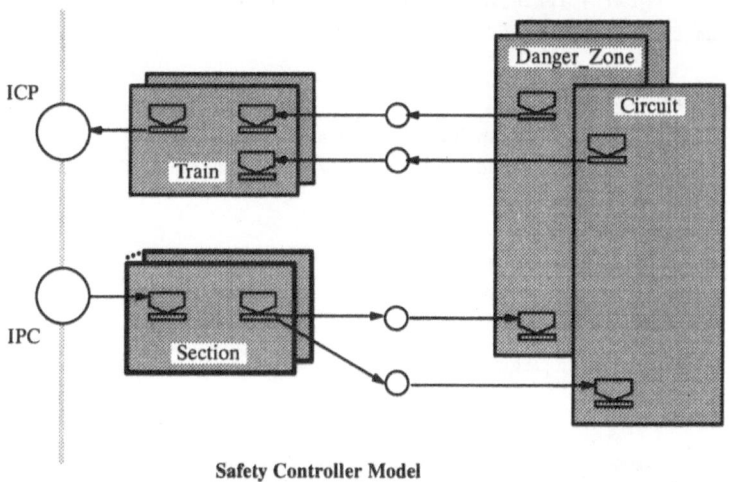

Safety Controller Model

Figure 4: The box structure of the PNO model of the safety controller

```
        Type_of_DZ: {a, b} ;
        DZ_Addresses: array [a, b] of Danger_Zone ;

ATTRIBUTE
        ms: Section_Number ;              // Unique section identifier.
        C: Circuit_Addresses ;
        DZ: DZ_Addresses ;
        mc: Type_of_Circuit ;             // This section's circuit type.
        sdt: boolean ;                    // Sensor detects train.
        D_DZA, D_DZB: integer ;           // Distances from danger zones $

OPERATION

        METHOD Read_Sensor ;

PROCESS
        cycle
                Read_Sensor ;
                C[mc].Train_in_Section(ms) ;
                DZ[a].Train_in_Section(mc,ms,D_DZA);
```

```
        DZ[b].Train_in_Section(mc,ms,D_DZB);
    end cycle;
```

SPECIFICATION
```
    IPC<sn> ;    // a train has entered section sn.
    SNh<dt> ;    // a sensor detects there is a train in its section.
```

Initial Conditions

$$SN^0 := \{< F >\};$$

// PNO shown in figure 5.

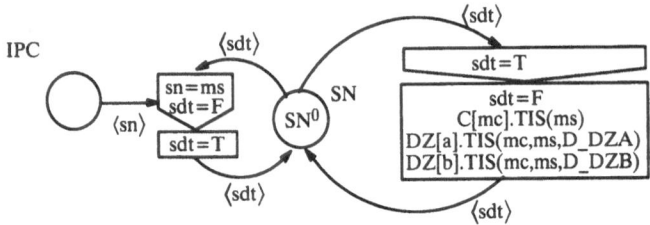

Figure 5: Specification of class Section

The specification for object **Circuit** that implements $RSS_{1,1,1}$, is presented below. When a sensor failure is detected an exception is raised and a handler is called to update the state of the safety controller to be consistent with the state of the physical process. If the number of consecutive sensor failures exceeds $mcsf$ then a handler will shut down the whole system because safety can no longer be guaranteed. A similar procedure is followed for an actuator failure. $RSS_{1,1,1}$ is implemented by **Localize_Sensor_Failures**, **Localize_Actuator_Failure** and **Test_Reserve_Section** which are specified in terms of the THL rules presented earlier.

CLASS Circuit

PUBLIC
```
        Train_in_Section;
```

```
TYPE
        Table_of_Trains: array [1,No_of_Trains] of Train ;
        Type_of_Circuit: {p,s} ;

ATTRIBUTE
        snj, snk: Section_Number ;
        tn: Train_Number ;
        mcsf: integer ;        // Maximum consecutive sensor failures.
        fsf: boolean ;
        T: Table_of_Trains ;
        mc: Type_of_Circuit ;

OPERATION

        METHOD Train_in_Section (IN sn: Section_Number) ;

        METHOD Localize_Actuator_Failure
                [ Failure : Shut_Down ] ;

            // RSS_{1,1,1} (rssd);


        METHOD Localize_Sensor_Failures
                [ Minor_Failure : Test_Reserve_Section ;
                Major_Failure : Shut_Down ] ;

            // RSS_{1,1,1} (rssc);



        METHOD Test_Reserve_Section;

            // RSS_{1,1,1} (rssa and rssb);


PROCESS
        cycle

                accept Train_in_Section ↦

                        Localize_Actuator_Failure;
                        Localize_Sensor_Failure;
                        Test_Reserve_Section;
```

```
                    T[tn].Move_Train_Section(mc,sn);
      end cycle;
```

SPECIFICATION
```
      NS<tn,snk,fsf> ;        // train tn is expected to enter section snk.
      RS1<tn,snj> ;           // train tn has reserved section snj.
      RS2<tn,snj>+<tn,snk> ;  // train tn has temporarily
                              // reserved snj and snk.
      FS<snk> ;      // section snk is not reserved.
      SS<tn,fsf> ;   // flag shows whether train tn has localized
                     // any sensor failures.
      TA<tn,snj> ;   // section snj was the last section
                     // occcupied by train tn.
      NOS<snj> ;     // snj is the new occupied section.
      STO<tn,snj> ;  // snj is the section to be occupied by train t$
```

Initial Conditions
```
          RS1^0 :={<1,3>,<1,2>,<1,1>,<1,0>,<2,11>,<2,10>,<2,9>,<2,8>} ;
          FS^0 :={<4>,<5>,<6>,<7>,<12>} ;
          TA^0 :={<1,3>,<2,11>} ;
          SS^0 :={<1,F>,<2,F>}
```

// PNO shown in figure 6.

The object instances of class **Train** stop a train within a section or allow a train to move to the next section. From the specification of the class **Train** presented below, we notice that a train is not allowed to move to the next section unless it has got permission from the object **Circuit**, and when a train moves into a danger zone, it must also have permission from the object **Danger_Zone**. The constants DZA and DZB in the PNO model represent the section numbers of circuit Cc that are crossing sections.

CLASS Train

PUBLIC
```
      Move_Train_Section;
      Move_Train_DZ;
```

TYPE

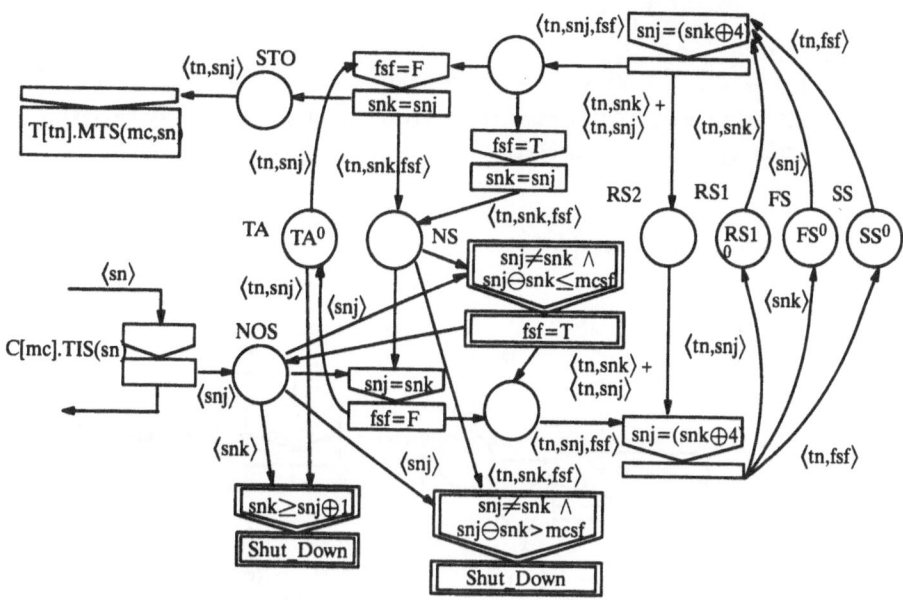

Figure 6: Specification of class Circuit

```
Type_of_Circuit : {p,s} ;
```

ATTRIBUTE

```
    tn: Train_Number ;
    sn: Section_Number ;
    DZA, DZB: Section_Number ;
    cn: Type_of_Circuit ;
```

OPERATION

```
    METHOD Move_Train_Section ( IN cn: Type_of_Circuit ; IN sn: integer) ;
    METHOD Move_Train_DZ ( IN cn: Type_of_Circuit ; IN sn : integer) ;
```

PROCESS

```
    cycle
        accept Move_Train_Section ;
        accept Move_Train_DZ ;
    end cycle;
```

SPECIFICATION
```
        ICP<tn,sn>;      // a train tn is allowed to enter in section sn.
        AES<cn,tn,sn>;   // in circuit cn train tn is allowed enter section sn.
        AED<hcn,tn,sn>;  // in circuit cn, train tn is allowed enter DZ.


        // PNO shown in figure 7.
```

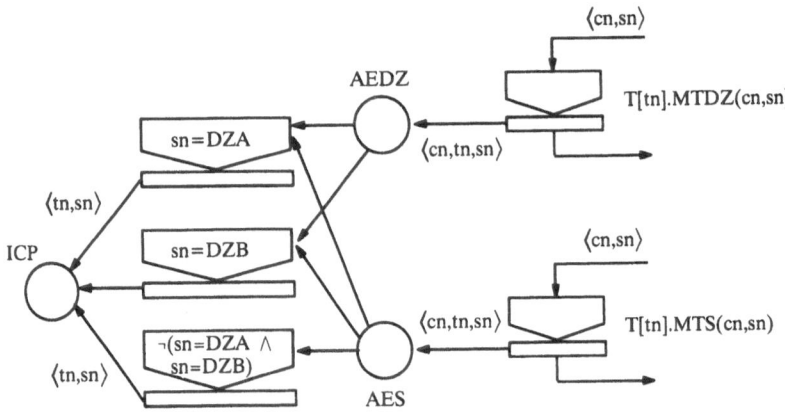

Figure 7: Specification of class Train

6.2   Controller Design

We now present a design for the safety controller software derived from
the specifications. In the interests of brevity, the details of the implementation of simple types (such as the types which represent places in
the PNO) are omitted. We will assume an implementation for such types
that allows tuples, in the form of records, to be added to or removed
from the array by simple arithmetic manipulation.

Design for Section

The object declaration given below shows a design for class **Section**
derived directly from the specification given above. The precise details of

basic type implementation have been omitted, but the mapping between specification and design can still be clearly seen.

```
OBJECT Section IS

ATTRIBUTE
        ms: Section_Number ;
        D_DZA, D_DZB: integer ;          // Distances.
        SN: boolean ;                    // = sdt.
        C : Circuit_Addresses ;
        DZ : DZ_Addresses ;
        mc : Type_of_Circuit ;

METHOD Read_Sensor
begin
        if <Signal from sensor> then
                SN := TRUE ;
end ;

PROCESS
        cycle
                Read_Sensor ;            // Sets value in SN
                if (SN) then
                begin
                        SN := FALSE ;    // Equivalent to destructive read.
                        C[mc].Train_in_Section(ms) ;
                        par
                                DZ['a'].Train_in_Section(mc, ms, D_DZA) ;
                        ||
                                DZ['b'].Train_in_Section(mc, ms, D_DZB) ;
                        end par ;
                end ;
        end cycle ;

END.
```

Design for Circuit

We now give the design for class **Circuit**. Owing to the complexity of this class and, once again in the interests of brevity, only part of the design is given, with the implementations of some of the methods (notably Localize_Actuator_Failure and Test_Reserve_Section) being omitted. However, the design for those methods which are not shown can be derived from the specification in the same way as that for those which are given.

```
OBJECT Circuit IS

ATTRIBUTE
    T: Table_of_Trains ;
    NOS: Set_of_Section_Numbers ;
    mcsf: integer ;
    FS: Table_of_Sections ;
    TA, RS1, RS2: Reservation_Table ;
    NS: Reservation_Flag_Table ;
    B1, B2: Reservation_Flag_Record ;
    SS: Train_Flag_Table ;
    mc: Type_of_Circuit ;
    Sensor_Fail: boolean ;

METHOD Train_in_Section (IN sn: Section_Number)
begin
    NOS := NOS + sn ;        // Adds sn to set NOS.
end ;

METHOD Localize_Sensor_Failure SIGNALS Minor_Failure, Major_Failure
begin
    ti: integer ;                    // Table index.
    No_Failure: boolean ;
    Current: integer ;
    No_Failure := FALSE ;
    ti := 0 ;
    Current := NOS.Get ;

    repeat
        ti := ti + 1 ;
        if Current = NS[ti].sn then
                begin
                    No_Failure := TRUE ;
                    B1 := B1 + New_Entry(NS[ti].tn, Current, FALSE) ;
```

```
                        NS := NS - NS[ti] ;
                end ;
    until (ti = NS.No_of_Entries) or No_Failure

    if No_Failure != TRUE then
    begin
            for ti := 1 to NS.No_of_Entries do

                if Current <= NS[ti].sn ⊕ mcsf then

                        begin
                            B1 := B1 + New_Entry(NS[ti].tn, Current,TRUE) ;
                            NS := NS - NS[ti] ;
                            signal Minor_Failure ;
                        end ;
            signal Major_Failure ; // This will only be signalled
                                   // if a minor failure has not already
                                   // been detected.
    end ;
end ;


PROCESS
    ti: integer ;
    Found: boolean ;
    Current: B2_Tuple ;

    cycle

        accept Train_in_Section ↦

            begin
                Localize_Actuator_Failure
                                    [ Failure: SHUTDOWN ] ;
                repeat
                    Sensor_Fail := FALSE ;
                    Localize_Sensor_Failure
                                    [ Minor: Sensor_Fail := TRUE
                                      Major: SHUTDOWN ] ;
                    Test_Reserve_Section ;
                until (Sensor_Fail = FALSE) ;
                Current := B2.Get ;
                NS := NS + Current ;
```

```
                    if Current.fsf = FALSE then
                          Trains[Current.tn].Move_Train_Section(mc, Current.sn) ;
            end ;
     end cycle ;


END.
```

Design for Train

Finally, we give the design for class **Train**. Like class **Section** this
is a simple object and it will be shown in its entirety (but for type
implementation details).

```
OBJECT Train IS


ATTRIBUTE
     AES, AEDZ: CTS_Record ;
     DZA, DZB: Section_Number ;
     tn: Train_Number ;


METHOD Move_Train_Section ( IN cn: Type_of_Circuit ; IN sn : Section_Number)
begin
     AES := New_Entry(cn, tn, sn) ;
end ;


METHOD Move_Train_DZ ( IN cn: Type_of_Circuit ; IN sn: Section_Number)
begin
     AEDZ := New_Entry(cn, tn, sn) ;
end ;


PROCESS
     cycle

          accept Move_Train_Section ↦

               begin
                        if AES.sn != DZA and AES.sn != DZB
                        then
                                AES := Empty ;
                        else
                                begin
```

```
                                   repeat
                                         accept Move_Train_DZ ;
                                   until (AEDZ = AES) ;
                                   AES := Empty ;
                                   AEDZ := Empty ;
                             end ;
                 end ;
            end cycle ;

END.
```

7   Conclusions

The basic aim of this paper was to present a systematic method to cope with exception handling during the specification and design of real-time software. The proposed method was described in the context of a framework developed for the requirements analysis of safety-critical systems. The method employs formal notations during specification, and an object-based approach for design. To illustrate how the proposed method copes with exceptions we considered exception handling for those failures in sensors and actuators which might have an adverse affect on the safety of the system. Guidelines for constructing a program from the formal specification were presented in the form of a mapping between the notations used during specification and design.

The approach to exception handling presented in this paper provides a framework which allows failures to be related to the specification of exceptions and facilitates the derivation of programs from the specification. This approach was shown to be feasible by applying it to a train set example.

Acknowledgements

Earlier discussions with Tom Anderson and Santosh Shrivastava helped in the development of this work. The authors would like to acknowledge the financial support of CAPES/Brazil, the Basic Research Action PDCS and BAe (DCSC).

References

[1] A. Saeed, R. de Lemos, and T. Anderson, "The Role of Formal Methods in the Requirements Analysis of Safety-Critical Systems: a Train Set Example," *Proc. of 21st Symp. on Fault-Tolerant Computing*, pp. 478-485, Montreal, Canada, 1991.

[2] F. Cristian, "Robust Data Type," *Acta Informatica*, vol. 7, pp. 365-397, 1982.

[3] M. Bidoit et al., "Exception Handling: Formal Specification and Systematic Program Construction," *IEEE Trans. on Software Eng.*, vol. 11, pp. 242-252, 1985.

[4] R. de Lemos, A. Saeed, and T. Anderson, "Analysis of Timeliness Requirements in Safety-Critical Systems," *Proc. of Symp. in Formal Techniques in Real-Time and Fault-Tolerant Systems*, ed. J. Vytopil, Lecture Notes in Computer Science 571, Springer-Verlag, pp. 171-192, Nijmegen, Netherlands, 1992.

[5] F. Jahanian and A. Mok, "Safety Analysis of Timing Properties in Real-Time Systems," *IEEE Trans. on Software Eng.*, vol. 12 , pp. 890-904, 1986.

[6] F. Cristian, "Exception Handling," *Dependability of Resilient Computers*, ed. T. Anderson, BSP Professional Books, Oxford, pp. 68-97, 1989.

[7] G. L. Wells, "Formal Safety Studies," *Safety in Process Plant Design*, John Wiley & Sons, New York, pp. 101-120, 1980.

[8] D. Powell, "Fault Assumptions and Assumption Coverage," *ESPRIT PDCS Second Year Report* vol. 1, chp. 5, 1991.

[9] A. Saeed, T. Anderson, and M. Koutny, "A Formal Model for Safety-Critical Computing Systems," *Proc. of SAFECOMP'90*, pp. 1-6, London, UK,1990.

[10] C. Sibertin-Blanc, "High-Level Petri Nets with Data Structure," *Proc. of 6th European Workshop on Application and Theory of Petri Nets*, Espoo, Finland, 1985.

[11] H. Genrich, "Predicate/Transition Nets," *Petri Nets: Central Models and their Properties*, eds. W. Brauer, W. Reisig, and G. Rozemberg, Lecture Notes in Computer Science 254, pp. 206-247, 1987.

[12] C. Ghezzi, D. Mandrioli, S. Morasca, and M. Pezzè , "A Unified High-Level Petri Net Formalism for Time-Critical Systems," *IEEE Trans. on Software Eng.*, vol. 17, pp. 160-172, 1991.

[13] R. de Lemos, A. Saeed, and T. Anderson, "A Train set as a Case Study for the Requirements Analysis of Safety-Critical Systems," *The Computer Journal*, vol. 35, pp. 30-40, 1992.

[14] S. K. Shrivastava, and A. Waterworth, "Using Objects and Actions to provide Fault-Tolerance in Distributed Real-Time Applications," *Proc. of Real-Time Systems Symposium*, pp. 276-285, San Antonio, Texas, 1991.

[15] S. Morasca, and M. Pezzè , "Validation of Concurrent ADA Programs using Symbolic Execution," *Proc. of 2nd European Software Engineering Conf.*, Lecture Notes in Computer Science 387, pp. 469-486, Coventry, UK, 1989.

[16] M. Combacau, and M. Courvoisier, "A Hierarchical and Modular Structure for F.M.S. Control and Monitoring," *Proc. of 1st Int. Conf. on AI, Simulation and Planning in High Autonomy Systems.* Tucson, AZ, 1990.

Realizing Changes of Operational Modes with a Pre Run-Time Scheduled Hard Real-Time System

G. FOHLER*

Institut für Technische Informatik

Technische Universität Wien

Treitlstr. 3/182

A-1040 Wien, Austria

email: gerhard@vmars.tuwien.ac.at

Abstract

Most hard real-time computer systems control processes that undergo several, mutually exclusive modes of operation. This paper discusses issues of handling mode changes and requirements for their application. Specification of mode changes, construction of static schedules for modes and transitions, and run-time execution of mode changes are presented. We propose concepts for mode changes in the context of MARS, a pre run-time scheduled hard real-time system. Our methods adhere closely to the ones established for single modes.

By decomposing the system into a set of disjoint modes, the design process and its comprehension are facilitated, testing efforts are reduced significantly, and solutions are enabled which do not exist if all system activities of all modes are combined into a single schedule.

Key Words: Distributed Hard Real-Time Systems, Pre Run-Time Scheduling, Precedence Constraints, Operational Modes, Mode Changes

---

*This work has been supported in part by ESPRIT Basic Research Action No. 3092 - Predictably Dependable Computer Systems and the Austrian Science Foundation (Fonds zur Förderung der wissenschaftlichen Forschung) under contract P8002-TEC.

1   Introduction

Processes controlled by hard real-time systems[1] typically exhibit mutual exclusive phases of operation and control. Such systems require a pre run-time verification, so that all timing constraints will be met. An aircraft control system, for example, performs different tasks during take off-, flight-, and landing phase. Heating up a vessel to operating temperature requires other control mechanisms, time bounds and even equipment than maintaining the normal setting. An imminent danger of overheating forces all activity other than emergency cooling to be canceled.

Such changes in the modes of operation affect the selection of control loops performed by the hard real-time system, their timing require-ments, attributes such as "critical" or "hard real-time", the set of acti-vated control programs, their dependencies, and execution times. Even major modifications in the computer system's structure such as its re-configuration and the change of the number of redundantly executing processors, are conceivable.

Several issues of hard real-time system design and operation are affected by the way different modes and changes between them are handled by the controlling system:

Solutions: Current scheduling methods hardly support different modes of operation adequately. Rather, all tasks carried out in different modes are handled as if they could occur at the same time, incorporating all computations into a single schedule. Although it is known which re-sources cannot be claimed at the same time, they have to be reserved as if they could. The price paid for not exploiting knowledge about mu-tually exclusive modes is an enormous overhead. Solutions which exist for each single mode are prohibited by combining all modes into a single one.

Design and Understandability: A design process that requires all modes of a system to be considered and designed at once, although they are logically independent, contradicts the principles of modularization and separation of concerns. It is also more difficult to understand which activities are coherent, which interactions are possible and relevant, and

---

[1]By *hard real-time system* we mean a system in which catastrophic consequences can be adjudged to be due to timing failures, as defined in [1]

in which mode specific actions are performed. Given the specification and the schedule of an "all-modes-in-one" system, it is a tedious task to separate these concerns, whereas they are evident for a set of different modes with individual designs and schedules.

Conventional systems necessitate the design process to take care of dependencies between logically separate parts. If, for example, a message m is sent exclusively in a mode $M0$, and message n in mode $M1$, they are logically independent and will never compete for transmission. In a "all-in-one" schedule they have to be dealt with as if they would. During schedule construction, giving transmission medium access to m affects n and vice versa in that case. They are no longer independent.

Testing: Instead of testing each mode and transition separately, all activities have to be tested at once in traditional systems, thus providing an input space that is dramatically larger than with mode separation.

For these reasons following separate modes of operation by assigning schedules to individual modes and transitions between them is of major concern. Whereas methods for constructing "single mode" schedules are well known, little consideration has been given to the changes between them. Sha et al. [2] have investigated mode changes for systems based on priority inheritance protocols.

Jahanian, Lee, and Mok presented Modechart, a graphical specification language for representing real-time systems [3]. The concept of modes in Modechart is that of fairly general design objects, which can be arranged hierarchically. A transition between modes represents a change in the control information of the system. It is an instantaneous event, taking no time. We view a mode as representing an operational phase which is performed by a single schedule. The modes comprising a system are mutually exclusive. The problem of transitions is that of switching between schedules performing different, mutually exclusive operational modes. So our main focus is on the algorithmical problems for scheduling.

This paper presents an approach to handle mode changes in the time-triggered, strictly periodic, pre run-time scheduled system MARS, *e.g.*,[4, 5].

The rest of the paper is organized as follows: Section 2 lists requirements for handling mode changes, which are a baseline for further discussions. The issues of specification of mode changes, off-line construc-

tion of schedules, and their run-time execution are dealt with in section
3. The paper is concluded with a summary and discussion of the current
status in section 4.

2   Requirements

Listed below are requirements for handling mode changes:

- Deterministic temporal behavior has to be achieved within the hard
  real-time environment.

- A prerequisite for this demand is a means to specify transitions in a
  way that allows timing constraints to be associated with transition
  definitions.

- Specification of mode changes has to be comprehensive, and adher-
  ent with the design of single modes. This allows designers to apply
  familiar methods, *e.g.,*precedence constraints and tools.

- In order to retain continuous system operation, activities executing
  during all modes have to operate without interference from the mode
  changes, *e.g.,*timing constraints have to be met. This requirement
  does not apply to emergency modes, such as system shutdown or
  establishing a stable, safe state.

- In addition to each mode, a verifiable timing behavior has to be
  achieved for mode changes, *i.e.,*it must be possible to prove before
  the system's run-time that all mode changes will be completed in
  time.

3   Concepts for Mode Change Treatment

This section presents approaches to specifying activities performing mode
changes, how adequate static schedules are constructed off-line, and their
run-time execution.

3.1   Specification

Precedence constraints are a commonly used means to express dependencies and execution orders between modules (called tasks further on) of programs. Tasks and their precedence constraints can be represented as acyclic, directed graphs, the so called precedence graphs. Nodes denote tasks and arcs the corresponding precedence relations. Often messages are assigned to arcs in the precedence graph. The precedence constraints of a task are fulfilled when the conditions associated with all edges leading to it have become true, *e.g.,*it has received all messages. The start node(s) (called stimulus tasks) make data of the environment available for further processing in the system, *e.g.,*by reading sensors. The end node(s) (called response tasks) transfer data to the environment, *e.g.,*by setting actuators. Deadlines can be associated with system activities expressed as precedence graphs in a straightforward manner. We call the time between the start of the first stimulus task and the completion of the last response task the maximal response time (MART) of a precedence graph.

Conventional precedence constraints allow only the definition of a single set of fixed dependencies on a fixed set of tasks for the whole system life time.

Keeping in mind that the system undergoes different modes of operation, we extend the concept of precedence constraints by a second dimension, the modes of operation. Practically speaking, we associate with each edge of the precedence graph the modes it applies in. The precedence constraints of a task T in mode M are fulfilled, when all edges to task T in mode M are fulfilled[2]. Note, that conventional precedence constraints can be viewed as a special case, with only one mode. The alteration of task sets, synchronization requirements, and deadlines can be expressed. Consider this very simple example: There are 9 tasks, TA, TB, $\ldots$, TI, three modes, MO, MT, MF and the following precedence relations (The edges of mode MO are depicted as solid lines, of MT as dashed, and MF as dotted ones) as illustrated in figure 1.

Tasks TA, and TC execute in modes MO and MT, TG in both MT and MF. The other tasks execute in single modes only. Alternatively, this can be denoted by labeling edges with associated modes, as for example

---

[2]More precisely speaking, when the the conditions associated to these edges are fulfilled.

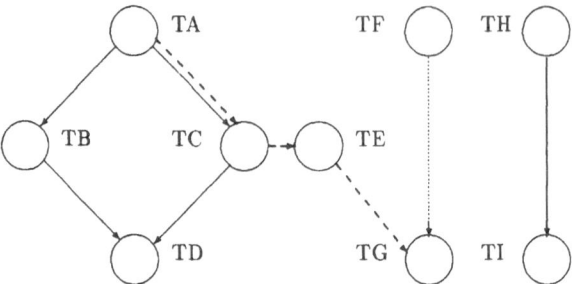

Figure 1: Example of Mode Change Specification by Two-Dimensional Precedence Constraints

$TA \xrightarrow{MO} TB$. Dependencies applying throughout several modes can be expressed by a list of modes: $TA \xrightarrow{MO,MF} TC$. An edge $M \xrightarrow{all} N$ indicates that $M \longrightarrow N$ will be performed in all modes.

We can now envision the system operation as follows: One mode is selected for starting, e.g.,doing initialization procedures. Thus the schedule created for this mode is executed periodically until a mode change is requested at time t. Then, the indicator for "current mode" is set to the new mode, and the execution is switched over immediately to the schedule of the new mode at the corresponding point of time t within its period[3]. The two schedules can be viewed as simultaneously running, but only the one whose mode is equal to the "current mode" indicator is actually performed. While acceptable for a principal mechanism, this method is not always satisfying, because it implements a "brutal" switch over, which does not care about fulfillment of precedence constraints or data consistency.

So we need to think about more elaborate, smoother forms of transitions. Again, we adhere to precedence constraints: Each transition is specified by a precedence graph. Such a graph can be defined on tasks of both the old and the new mode. A mode is assigned to the corresponding transition precedence graph as well.

In the example, MT (with dashed line edges) is such a mode constituted by a transition precedence graph: $TA \rightarrow TC \rightarrow TE \rightarrow TG$. Data read and processed by TA and TC are used by the transition for an initial

---

[3]That is t modulo period, when schedules are executed periodically.

input to TG, after being adapted by TE. A state message concept as applied in MARS facilitates this "data recycling": Task TC writes its output data (a message) to some buffer, from where it can be read by TD or TE.

Transition precedence graphs are handled slightly differently at runtime: In contrast to graphs of continous modes, which execute periodically until a mode change is initiated, they execute only once. Note that precedence graphs executing in both the old and the new mode are part of the transition schedule as well.

Mode Change Delays and Deadlines: We define the time bound of a mode change from the point of view of the environment: The deadline of a mode change is the time interval between the start of the last stimulus task before the mode change request in the original mode and the completion of the last response task before the new mode is established. This deadline is a requirement in addition to the individual deadlines of the precedence graphs. Taking into account the run-time mechanisms described in section 3.3, the off-line scheduler has to secure that these deadlines are kept, as well as the individual ones. If it is unable to do so, the system designer has to take appropriate actions, maybe resulting in a redesign of the system.

Semantic Constraints: Particular attention has to be given to the specification of how ongoing activities are handled when a mode change is requested. There are three possibilities, as pointed out by Jahanian et al. in [3]:

- An *immediate change, aborting current activities* is specified by a transition precedence graph consisting of no task, an empty one. Since the schedule of the new mode starts executing once all tasks of the transition precedence graph are done, an empty graph models immediate switch over.

- The graph requiring *completing all current activities before changing*, the other extreme, is identical to the graph of the old mode. Since the originating and the transition precedence graphs are the same then, switching to the transition, which can take place instantly, does not change the current schedule execution. Only after it is completed, the change to the destination mode schedule is performed.

- *Completing some of the current activities* is specified by creating an transition precedence graph consisting of the desired activities of the mode old plus maybe new ones. Thus the transition can be designed to process results computed in the old schedule and use them for preparing the new mode.

A mode change requested while a transition is executed is defined in the same way as occuring in a continous mode: A transition precedence graph with one of the semantics given is specified.

Changing from one operational phase of the system to another may involve a number of transition schedules, depending on the activities needed to establish the new mode. A vessel control system for example, manages the change from filling to heating up. It changes from filling mode to a mode in which safety requirements are checked, a temperature increase plan depended on the quality and amount of material filled is calculated, and checks are performed whether this is according to efficiency standards. Finally it initializes heating devices, and once they are ready, changes to the heating mode.

Design at Mode-Level: Speaking of modes and transitions, the idea of designing all mode changes by specifying a state machine is at hand. A design system will allow the designer to create this machine first, without considering details of the modes and transitions involved. The designer is then forced to specify precedence graphs with their timing constraints for all modes and transitions, thus supporting completeness of the design automatically. The same state machine will also provide complete information about possible transitions for testing. Resolution of conflicting mode change requests at run-time is specified by using information provided by this machine. We are currently investigating this topic in more detail.

Maintaining a consistent global view on the mode change activities is a non trivial task. It is supported by the reliable multicast communication services in MARS [6].

3.2 Constructing Schedules

We present an approach to off line scheduling of mode changes on the basis of the MARS pre run-time scheduler [7] [4]: Given a set of precedence graphs with tasks allocated to components, the algorithm attempts to find a feasible schedule by appliying a heuristic search.

It traverses precedence graphs along their edges and tries to select the set of tasks out of those with fulfilled precedence constraints, called ready tasks, that seem to be the most promising step towards a feasible schedule at predefined times. These selections are called scheduling decisions, and have the form "At time 5, out of tasks A, B, C, D select A to execute".

In other words, it constructs schedules by assigning tasks to CPU-slots successively according to the specified precedence order. For the precedence graph of mode MO in the example, successive steps could be as given in figure 2.

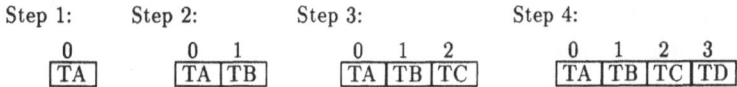

Figure 2: Example of Construction Steps for a Single Mode Schedule

The basic idea for our mode change scheduling approach is similar to that for two dimensional precedence constraints: We assume – for pre run-time scheduling – the existence of one cluster per mode (called virtual cluster), including transition modes. Instead of traversing the precedence graphs to search for a single schedule, we traverse all graphs of all modes at the same time, trying to construct an individual schedule for each mode. If a task has to execute in more than one mode, it has to be scheduled in the same way in all virtual clusters, *i.e.*,at the same time. This enables switching between schedules at run-time without rephasing. Scheduling decisions do no longer select tasks out of a set of a single mode, but out of sets of tasks for each mode together at once, in accordance with the just explained "switch through" requirement. These decisions have the form "At time 5, out of tasks A, B, D select

---

[4]For reasons of clarity, we restrict the following explanations to tasks executing on a single processing unit (called component) only. Both the single mode and the mode change scheduler work for a number of multiple components, a so-called cluster.

A to execute in mode M0, and out of tasks E, C select C in mode M1".
The rest of the search control mechanisms remain almost unchanged.
In the case of the example the performed steps may be given in figure
3 (tasks that are "switch through" are connected by arrows between
modes). As with the base algorithm, a pre run-time verification of the

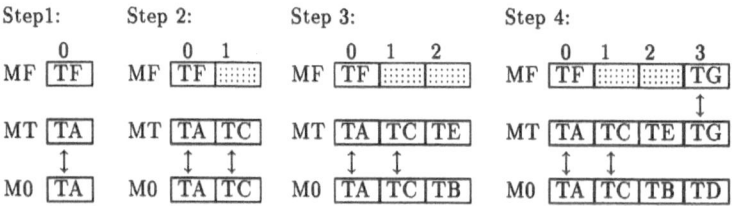

Figure 3: Example of Construction Steps for a Mode Change Schedule

timing behaviour is possible.
We have implemented a prototype version of the off-line scheduler. Using this, we want to find out more about limitations and problems with
different kinds of task sets and modes. Based on these results we want
to revisit these concepts and improve them accordingly.

3.3   Changing Modes at Run-Time

We are now able to propose the following method for changing modes:
The schedule of a continous mode is executed periodically until a mode
change is requested at time t. Then it is checked, if discontinuing the
current schedule and continuing immediately with the transition schedule – according to the pre calculation described in 3.2 – is feasible. This
is true when canceling all activities of the current mode frees enough
CPU-time to accommodate the new activities of the transition schedule. In this case data produced by the current schedule so far are further
processed by the transition schedule. Otherwise the transition schedule
will be switched after the current schedule has completed execution.
The maximal delay is considered in the construction of the schedules[5] .
Based on knowledge of resource needs of schedules of the modes, a pre
run-time analysis calculates a point in time within the schedules' period,

---

[5]Since the off-line scheduler bases on a heuristic search strategy, this can be incorporated fairly easily by adding another feasibility constraint

up to which the transition can be accommodated. The mentioned run-time test checks whether current time is smaller than that point. The "black-out" time between this point and the start of the next period contributing to the transition time has to be considered during the construction of the schedules. Once the transition is completed, scheduling is switched to the new mode. We will illustrate this mechanism using the example of section 3.1. The reaction to the data read by TA (*i.e.,*the time between TA and TD, and TA and TG) has to be done within 7 slots. TA reads data every 7 slots. The deadline for the mode change (*i.e.,*from the last start of TA till the first completion of TG) is 10 slots. For simplicity of this example, each task requires one CPU slot for its execution. The pre-calculated schedules for periodic execution are:

	0	1	2	3	4	5	6
M0	TA	TC	TH	TB	TI	TD	░░░

	0	1	2	3
MF	TF	░░░	░░░	TG

The transition schedule for MT is

	0	1	2	3
MT	TA	TC	TE	TG

The transition needs to execute two tasks, TE and TG, that are not in MO. These can be accommodated for mode change request times till the end of slot 4. The execution of the graph $TH \rightarrow TI$ can be aborted abruptly, since it is not part of the transition precedence graph[6] If a special treatment for the completion of $TH \rightarrow TI$ is required, this has to be specified in the transition precedence graph.

Let us assume a request after the end of slot 2. The transition can be accommodated, since 2 is smaller than 4. In MO, tasks TA, TC, and TH have been completed. Since the transition is designed to make use of the results produced by TA and TC, it may start with TE immediately. After the completion of TG, the schedule for MF starts to execute periodically. The resulting execution order of this mode change is then

0	1	2	3	4	5	6	7	8
TA	TC	TH	TE	TG	TF	░░░	░░░	TG

With these schedules for MO and MT, the mode change deadline is met for requests at any time, even during the "black-out" time in slots

---

[6]It is conceivable that TH and TI may be required to complete execution, *e.g.,*because of their inner state.

5 and 6. Consider a request at time 5: The next instance of task TA starts at slot 7, since its period has to be retained. Then the transition schedule is executed, which is completed by slot 10. All precedence graph deadlines, TA's period and the mode change deadline are met.

We have implemented a prototype version of a simulator for the run-time execution of mode changes.

4   Summary and Status

In this paper we discussed the necessity of adequatly handling systems with different modes of operation, and listed requirements for their proper treatment. The conventional approach of combining system activities of all modes into a single schedule results in an overhead prohibiting solutions, which exist if the system is decomposed into a set of single modes. This common approach contradicts the design principle of modularization, reduces understandability of the designed system's activities, and introduces dependencies between logically independent activities. Testing is complicated as well.

We presented a method for expressing and scheduling mode changes for the MARS system by extending the concept of precedence constraints. When a mode change is initiated at system run-time, it is accomplished by switching through a series of precalculated transition schedules, that prepare for the new mode.

We provided a basis for scheduling mode changes that allows the pre run-time scheduling algorithm for single modes to be extended for this purpose.

Our concepts adhere to the design principles established so far for single mode treatment, and meet the requirements given for mode changes. Most importantly, mode changes are designed for deterministic, verified temporal behavior and are executed accordingly.

We have implemented the pre run-time scheduling extension as described in section 3.2 and a simulation environment. Experimental evaluation of the presented methods and aspects of incorporating them into MARS, such as operating system support and integration in the design environment, will be main areas of future research. We will also investigate issues of origin and initiation of mode change requests, and derivation of timing constraints for the mode changes.

Acknowledgements

The author wishes to thank Krithi Ramamritham for fruitful discussions and helpful comments, the members of the MARS research team, in particular Peter Puschner, Günter Grünsteidl, and Hannes Reisinger for establishing a creative atmosphere and countless discussions, and the referees for competent and relevant remarks. The students Peter Hackl and Martin Pottendorfer implemented the prototype mode change scheduler and the run-time simulation environment.

References

[1] A. Vrchoticky and W. S. (Eds.), "Report on the first pdcs workshop on real-time systems," *Research Report 3/90*, Institut für Technische Informatik, Technische Universität Wien, Vienna, Austria, January 1990.

[2] L. Sha, R. Rajkumar, J. Lehoczky, and K. Ramamritham, "Mode change protocols for priority-driven preemptive scheduling," *Real-Time Systems*, vol. 1, pp. 243–265, December 1989.

[3] F. Jahanian, R. Lee, and A. Mok, "Semantics of modechart in real time logic," *Proc. of 21st Hawaii International Conference on Systems Sciences*, pp. 479–489, Jan. 1988.

[4] H. Kopetz, A. Damm, C. Koza, M. Mulazzani, W. Schwabl, C. Senft, and R. Zainlinger, "Distributed fault-tolerant real-time systems: The MARS Approach," *IEEE Micro*, vol. 9, pp. 25–40, Feb. 1989.

[5] H. Kopetz, R. Zainlinger, G. Fohler, H. Kantz, P. Puschner, and W. Schütz, "An engineering approach to hard real-time system design," *Proc. of the Third European Software Engineering Conference, ESEC '91*, (Milano, Italy), pp. 166–188, Oct. 1991.

[6] G. Grünsteidl and H. Kopetz, "A reliable multicast protocol for distributed real-time systems," *Proc. of 8th IEEE Workshop on Real-Time Operating Systems and Software*, (Atlanta, GA, USA), pp. 19–24, May 1991. Pergamon Press, 1992.

[7] G. Fohler and C. Koza, "Heuristic scheduling for distributed real-time systems," *Research Report 6/89*, Institut für Technische Informatik, Technische Universität Wien, Vienna, Austria, April 1989.

Formal Specification and Simulation of a Real-Time Concurrency Control Protocol

P. V. D. STOK, L. SOMERS, P. THIJSSEN

Dept. of Math and Comp. Science

Eindhoven University of Technology

Postbox 513

NL-5600 MB Eindhoven, Netherlands

Abstract

The executable specification tool ExSpect is presented. A concurrency algorithm for a real-time distributed system is specified in the language of this tool. Parts of the specification are shown and it is demonstrated how the executable specification can assist in the concurrency control development.

1   Introduction

The DEpendable Distributed Operating System (DEDOS) is currently designed and implemented at the Eindhoven University of Technology (EUT) [1]. It is intended for the development and testing of different algorithms and control architectures. The system is targeted to the control of industrial processes and other physical processes (e.g. photocopiers, planes, production lines and physics experiments). The distribution is required for three purposes: (1) the process to control is distributed over a large area, (2) parallelism can be used to increase the efficiency of the control system and (3) the distribution allows the implementation of fault tolerance techniques. Two aspects of dependability as defined by Laprie [2] are taken into consideration: timeliness and reliability. The reliability of the control system is increased by replicating hardware resources. Two fault tolerance techniques are used: (1) the same versions of the software are executed concurrently on different hardware platforms to guarantee that the implementation meets the functional and timeliness specifications even when a number of specified hardware failures occur, or (2) the software is migrated from the failing hardware platform and restarted from some well defined state to meet the functional specification.

Two classes of timeliness specifications can be met by DEDOS. For that reason two types of executions can be defined: (1) Hard Real-Time (HRT) executions and (2) Soft Real- Time (SRT) executions. The first is used for the execution of periodic and Sparse Aperiodic (SA) executions of which the time characteristics are exactly known. They form the backbone of the control system and when a deadline of a HRT execution is not met, catastrophic consequences may result. HRT executions are used for closed loop controls and surveillance of critical equipment. HRT executions are scheduled off-line, and the precalculated schedule is executed by DEDOS on a repetitive basis [3]. Alternatives are allowed in the code, but all possible executions are known beforehand. Inside the off-line calculated schedule one or more gaps are reserved for the execution of SA executions. All SA executions are known beforehand and their execution frequency and execution time has an upper bound.

The timing constraints of the SRT executions are unknown. Their maximum executions times are known, but not their execution frequencies and their starting moments. In contrast to HRT executions, new SRT executions may be added and old ones removed. SRT executions are used for operator interaction and non-critical alarm visualization. When a SRT execution does not meet its deadline and is aborted, no catastrophic consequences occur but only an acceptable performance degradation results.

The DEDOS development is based on the object oriented paradigm [4]. The exchange of data between HRT and SRT executions takes place via objects [5]. HRT executions can modify data in Hard-Write (HW)-objects, which are later read by both SRT and HRT executions. SRT executions can modify data in Soft-Write (SW)-objects, which are read by HRT and SRT executions. An aspect of objects in DEDOS is the guarantee of exception and concurrency atomicity [6]. Exception atomicity means that all modifications on a set of related objects are executed completely or not at all; concurrency atomicity means that when two or more sets of modifications on a set of related objects are executed concurrently, the results of these modifications are equal to some serial execution of these sets. The atomicity specification does not guarantee that the results of the actions are not lost when memory or CPU failures occur. The part of DEDOS that handles fault tolerance and permanency of results is outside the scope of this paper.

An algorithm has been developed such that the atomicity of the objects can be guaranteed even when they are used by both HRT and SRT executions, while a SRT execution may never jeopardize the HRT executions. HRT executions may jeopardize and thus abort SRT executions, but the atomicity of the modifications on the objects remains guaranteed. The algorithm is based on Multi Version Time-Stamp Ordering (MVTSO) [7].

Two problems have to be solved, before a final implementation of the concurrency control algorithm on target hardware can be envisaged: (1) Is the chosen method not only correct but also viable and (2) is it possible to estimate for an application the number of versions that is required per object. At the EUT an EXecutable SPECification Tool (ExSpect) has been developed to solve exactly these types of problems [8]. A formal specification of a distributed algorithm based on a functional language and Petri nets can be executed to verify its functional and temporal aspects. In the rest of this paper an introduction of ExSpect is presented, followed by the description and specification of a part of the concurrency algorithm in ExSpect. The results obtained with the execution of the specification are described and the viability of the algorithm is discussed. The conclusions discuss the experience gained with ExSpect.

2   ExSpect Description

ExSpect (executable specification tool) is a formalism to describe distributed systems [8]. It is based on timed hierarchical colored Petri nets. (For colored Petri nets, see e.g. [9].)

Basically, the description of a system in the ExSpect formalism consists of two kinds of objects: channels (places) and processors (transitions). A channel is passive, it serves as a place holder of so called tokens: at any moment in time it contains an amount of these tokens. A channel can serve as input or output channel for a processor, see for example fig. 1. A processor is active: if all input channels of a processor contain at least one token, the processor may fire. If a processor fires, it removes

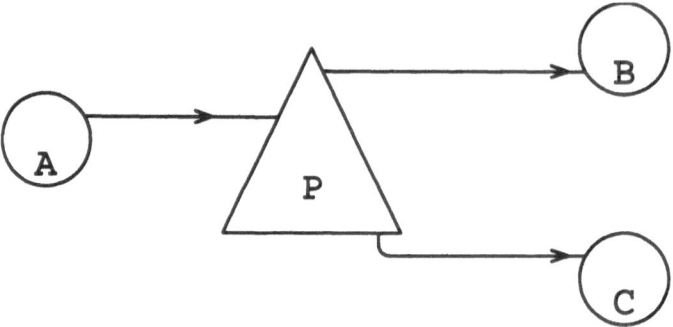

Figure 1: A processor (P) with input channel A and output channels B and C

(consumes) a token from each of its input channels and produces tokens for its output channels. Each channel has a type that specifies which values (colors) are allowed for the tokens in that channel. The values of the tokens that are produced when a processor fires depend upon the values of the tokens that are consumed. Thus, a processor can be seen as a function from the types of its input channels to the types of its output channels.

A processor may have a precondition: only token values that obey this precondition will be consumed.

Tokens also carry a time stamp. A token may not be consumed before the (simulated) time is greater than or equal to its time stamp. In the processor a delay is specified for each produced token. The time stamp is equal to the time of firing plus the delay. The example in fig. 2 represents a processor that consumes one numerical token from channel A at a time. If this token is positive then a token is produced for channel B. Its value is a pair consisting of the incremented value of the consumed token and the string 'ok'. The time stamp of this token will be equal to the time of firing plus 1.5. If the consumed token is not positive, a token is produced for channel C without delay.

```
proc P [in A: num, out B: num   str, C: str] :=
if A > 0 then
        B ← <<A+1,'ok'>> delay 1.5
else
        C ← 'invalid'
fi
```

Figure 2: Specification of processor P

```
type X := [a:num, b:$str, c:str];
type pid := num;
type CPUstat := pid → [tm:real,prs:bool];
```

Figure 3: Some type definitions

The colors of the tokens may represent very complex data. The ExSpect type system has a few basic types and type constructors to build more complicated types. Among the basic types are the booleans (bool), rational numbers (num), floating point numbers (real) and strings (str). The type constructors allow for the construction of Cartesian products, sets, finite mappings (sets of pairs with unique first component) and tuples. Functions operating upon those types can be constructed with the help of a set of predefined functions. Examples of predefined functions are numerical functions like + and -, logical functions like "or", and tuple functions like @ for projection on a label of a tuple.

In fig. 3 some type definitions are given. Type X represents a tuple type with labels a, b and c. A set of strings ($str) is assigned to label b. Type CPUstat is a mapping from pid (a rational number) to a tuple with fields of type real and bool. An example of a tuple of type X is:

[a:123, b:{'aa','abc'}, c:'xyz'] .

A tuple may be projected on a label by means of the function @. For example,

[a:123, b:{'aa','abc'}, c:'xyz'] @ a

is equal to 123. A mapping is a set of pairs, so an example of a value of
type CPUstat is:

  {<<1,[tm:1.75,prs:true]>>, <<2,[tm:3.68,prs:false]>>,
  <<3,[tm:1.23,prs:true]>>}.

A mapping may be applied to a value of its domain by means of the
operator ".". The domain of the above example is {1,2,3}. Applying
this example mapping to 2 yields:

  [tm:3.68,prs:false] .

The function upd(x,y,z) updates a mapping x in point y with value z.
So, applying upd to the above mapping of type CPUstat with 2 for y
and [tm:4.56,prs:true] for z gives

  {<<1,[tm:1.75,prs:true]>>, <<2,[tm:4.56,prs:true]>>,
  <<3,[tm:1.23,prs:true]>>}.

Models of large systems will become very soon too complex to understand if they consist of only channels and processors. The hierarchy
concept solves this problem. Channels and processors may be grouped
into systems, such systems may be used in other systems and so on.

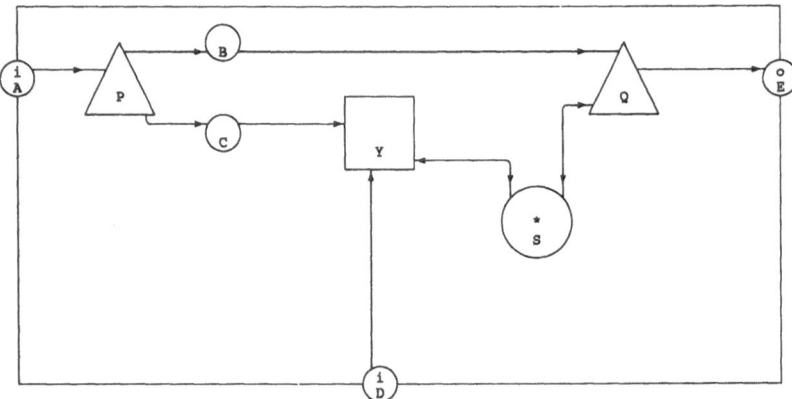

Figure 4: A system containing a store S and another system Y

The store is another concept that makes a model more compact. It represents a channel that is used as input and output by processors and contains always exactly one token. Such a token can be of a set or mapping type; thus, it may be considered as a database that is read and rewritten when a processor fires.

In Figure 4 a system is shown that consists of two processors (P and Q) and another system (Y). It receives tokens from its environment via channels A and D and outputs tokens to the environment via E. Store S is shared by system Y and processor Q.

3 Protocol Description (Informal)

The concurrency algorithm is designed for two types of objects: replicated objects and nested objects. Nested objects invoke methods of other objects. Replicated objects are identical and a modification on one object is executed on all or on none.

In this section, only the protocol for replicated objects is presented and some aspects of its specification in ExSpect are shown. Replicated objects are used for the realization of distributed shared data. Each application can access these data as if they are shared data supported by a mono-processor system. Such a construct simplifies the writing of application programs [10]. The concurrency and exception atomicity are guaranteed by using a modified version of the MVTSO protocol. Modifications are needed with respect to existing protocols for two reasons: (1) The number of versions per object is limited, and (2) SRT executions should in no way perturb HRT executions.

The first reason has an influence on the reading of the object's data. When a SRT execution reads a version of an object, it can be preempted and versions can be modified during its preemption. The finite number of versions has as consequence that the version which is currently being read can be overwritten, thus invalidating the data. Consequently, the SRT execution verifies after the read operation that its version has not been changed. This can be done by verifying the write Time-Stamp associated with each version. (N.B. The Time- Stamp of a version is a different entity from the time stamp attributed to ExSpect tokens.) When the write Time-Stamp has been modified, the read operation has to be restarted. A time- out is associated with each read operation.

When the elapsed time exceeds the time-out the execution is aborted. By counting the number of restarts, ok's and aborts during the simulation a good indication can be obtained about the number of versions per object that is required for a given application.

The second reason of non-perturbation of HRT executions has a profound influence on the structure of the protocol. To minimize the interference between the SRT and the HRT executions, objects are separated in two sets: modifiable by SRT or modifiable by HRT executions. The writing of objects by HRT executions is executed in a correct order calculated by the off-line scheduler. The writing of objects by SRT executions is delegated to a set of SA executions. The SRT execution sends a request for a SA execution to a request manager. In front of each gap in the schedule destined for the SA executions, the request manager is activated on all processors (CPU). One of the request managers orders all requests and distributes this ordering to all participating request managers. Each request manager now has the same order of aperiodic requests and activates aperiodic executions to execute these requests.

All SRT write requests will be performed on the SW-objects by the SA executions at the same time. Consequently, all HRT executions will read the same data when activated afterwards. The simulation will show the best strategy for communicating the SRT requests to the managers and for selecting the distributing request manager. Further, the simulation should again determine how many versions are required.

4 Protocol Description (Formal)

In Figure 5 the hierarchy of systems that constitute a simplified specification is shown. A more detailed specification is presented in [11]. At the highest level two systems, *admin* for the administration and *main* for the simulation, are defined. Via the channels *ok*, *abort* and *error* the corresponding results of the individual *request*'s are communicated to *admin* which maintains the percent of successes and failures of the requests in the store *statistics*. System *main* is constituted of four systems and one store. The store *objects* contains the description of all objects, their locations, their versions and their Time-Stamps. The system *concurrency- control* executes the requests on the objects as specified by the systems *aperiodic*, *hrt* and *srt*. The system *hrt* generates the

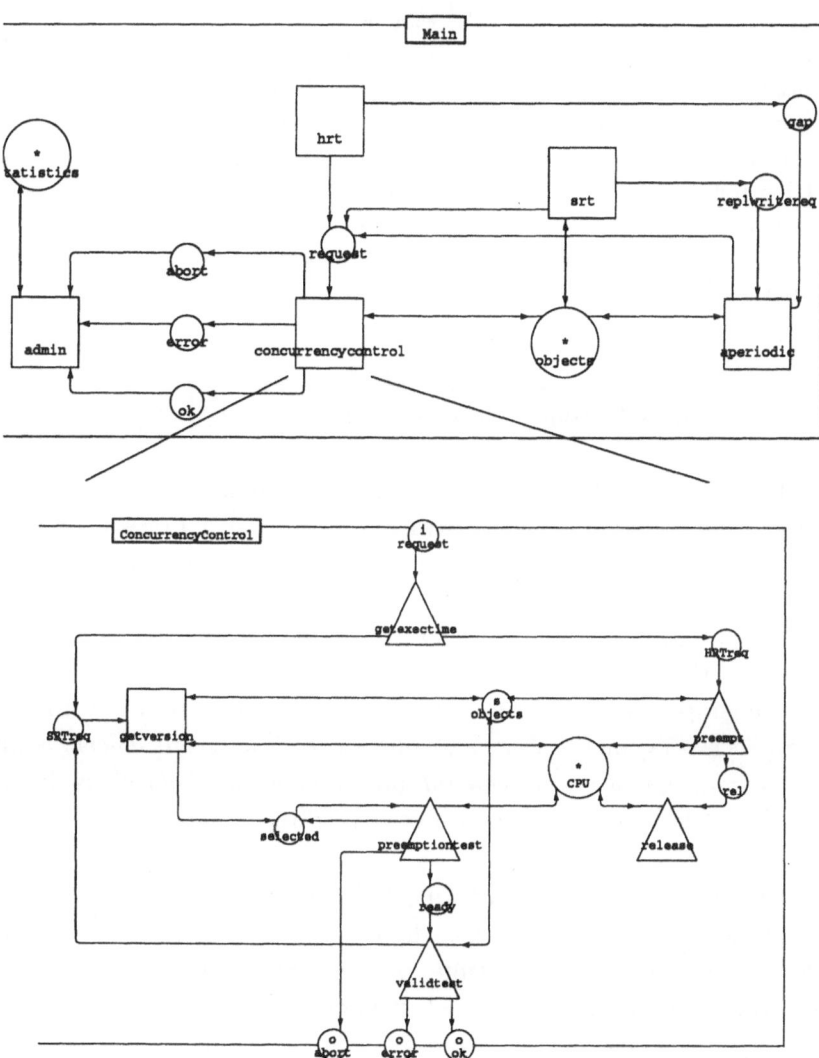

Figure 5: Hierarchy of systems for concurrency control specification

```
type pid := num;                        – CPU id
type mexec := real;                     – method execution time
type cid := num;                        – class id
type mid := [class:cid, tim:mexec];     – method id
type timestamp := real × pid;           – time with CPU id
type rversions := $timestamp;           – versions identified by Time-Stamps
type replobj := cid × $pid × rversions; – replicated objects
type TO := real;                        – Time Out
type request := [method:mid,cpu:pid,ts:timestamp,to:TO];
                                        – request message
type CPUstat := pid → [tm:real,prs:bool]; – CPU status
```

Figure 6: Representation of objects in ExSpect

requests originating from the off-line calculated schedule as specified by
the user. The system *srt* generates requests with a Poisson distribution of the intervals between the requests. Which request is executed
at these times is completely random. The system *aperiodic* is activated
each time a gap is met in the HRT schedule. It generates requests as
dictated by the request queue which is constructed from the write requests generated by *srt*. The system *concurrency-control* is composed
of five processors, one system and one store. The store *objects* is also
accessed from *concurrency -control*, but to show that *objects* is defined
at a different level it is marked differently with a character "s" instead
of a character "*".

The ExSpect representation of DEDOS concepts such as classes, objects, methods, CPU's is shown in fig 6. A method is uniquely defined
by its class and its execution time. A Time-Stamp is uniquely defined
by the time and the CPU in which it is created. Per object a number
of versions is defined by the set of write-Time-Stamps of these versions.
A replicated object is defined by its class, the CPU's on which it resides
and a number of versions for each object. The request is the entity which
moves through the simulation system and represents the request on a set
of replicated objects. These definitions represent the basic entities on
which the ExSpect functions act. The number of versions, objects, and
CPU's are defined by the user as prescribed by its application. The to-

```
proc preempt[in HRTreq: request, out rel: request, store CPU: CPUstat] :=
if (CPU.HRTreq@cpu)@prs then
      CPU ← upd(CPU, HRTreq@cpu,
        [tm:(CPU.(HRTreq@cpu))@tm+HRTreq@method@tim,prs:true]
else
      CPU ← upd(CPU, HRTreq@cpu,[tm:0.0, prs:true]),
      rel ← HRTreq delay HRTreq@method@tim
fi;
```

Figure 7: Specification of preempt.

ken (*request*) flow in the system *concurrency-control* will be discussed in more detail. The store CPU of type CPUstat contains for all CPU's the status (CPU is occupied or not) and the preemption-time of the SRT request. Two types of requests arrive via the input channel *request* and are treated by the processor *getexectime*. SRT requests with Time-Stamp zero are sent to *getversion* via the channel *SRTreq*. HRT requests with non-zero Time-Stamps are sent to *preempt*. *Preempt* adds the execution time of the request to the preemption time when the CPU in store *CPU* is not free. In fig. 7 a simplified specification of the treatment of a read request by *preempt* is shown. In the header the input and output channel plus the required store are specified. In the second line the method execution time specified in *request* is added to the preemption time of the CPU, if present in store *CPU*. If not present the specified CPU is replaced in *CPU* with preemption time zero; afterwards the request is passed to the channel *rel* with a delay specified by the execution time of the request. Processor *release* is activated after this delay and liberates the CPU.

The SRT requests are treated by the system *getversion*. When the required CPU in store *CPU* is free, a Time-Stamp is created and the version is selected in the store *objects*. A token with a delay equal to the execution time of the SRT request is sent over the channel *selected*. The preemption of the request by a HRT request is visible in store *CPU*. In case of preemption the request is reinserted into *selected* with a delay equal to the preempted time. However, when the time that passed since the start of the request is larger than its time-out, the request is not reinserted into *selected* but a token is set on the channel *abort*. The processor validtest verifies the overwriting of the selected version by comparing the Time- Stamp of the selected version with its own Time-Stamp. When overwriting has occurred, the request is represented to *getversion*; simultaneously, the request is sent to *admin* via the channel *error*.

5   Results

First results have been obtained for a configuration of twelve CPU's. The example described below can be thought typical of a not too complex piece of software, where a set of two CPU's monitors the control process executed by a set of ten other CPU's.

Two CPU's support only SRT executions which read or write data situated on the other ten CPU's, which in turn support both HRT and SRT executions. A schedule has been specified for twelve HRT executions. In one period one HRT execution writes to a set of replicated HW-objects situated on all ten processors, one HRT execution reads simultaneously on ten processors from one set of replicated SW-Objects, and ten HRT executions write and/or read nested objects distributed over all ten CPU's. A set of fifteen SRT executions read or write three nested objects also accessed by above mentioned HRT executions, to provide a background load of SRT executions. One SRT execution reads from a set of ten nested objects distributed over ten different CPU's. On the two separate CPU's, two SRT executions with an execution time equal to the period of the schedule read or modify a set of nested HW- objects or SW-objects also accessed by above mentioned HRT executions.

The timeout of the SRT executions is set equal to their execution time times a fixed number (called TOmlt). The number of versions is set equal to two as determined by the relation (longest SRT execution time)/(HRT period). The number of versions needed is roughly equal to the number of times a HW-object is modified during the longest SRT execution. This rule of thumb holds for not too complex nested object configurations. When SRT executions have to wait for other SRT executions before they can terminate their write operations, the picture changes dramatically. The simulation is essential to determine the number of versions required per object.

Table 1: Number of successes (Nsuc), Aborts (Nab) and retries (Nret) as function of the parameters Ntot, Rtot and TOmlt. No concurrency control is used.

Tomlt	Rtot	Ntot	Nab	Nret	Nsuc
10	309	306	36	0	575
10	278	150	18	0	408
10	112	310	21	0	399
20	309	306	14	0	593
20	278	150	4	0	423
20	112	310	9	0	412

Table 2: Number of successes (Nsuc), Aborts (Nab) and retries (Nret) as function of the parameters Ntot, Rtot and TOmlt. Wait with request for gap manager strategy is used.

Tomlt	Rtot	Ntot	Nab	Nret	Nsuc
10	309	306	258	12	283
10	278	150	120	22	242
10	112	310	249	38	142
20	309	306	238	22	300
20	278	150	111	33	249
20	112	310	240	57	147

Table 3: Number of successes (Nsuc), Aborts (Nab) and retries (Nret) as function of the parameters Nint, Rint and TOmlt. Send request to gap manager strategy is used.

Tomlt	Rtot	Ntot	Nab	Nret	Nsuc
10	309	306	226	19	382
10	278	150	90	33	336
10	112	310	219	41	199
20	309	306	205	47	397
20	278	150	86	48	338
20	112	310	209	58	205

The load of the HRT schedule is kept constant. The HRT load on the most heavily loaded CPU is 93%; the HRT load of the lightest loaded CPU is 2%. The load of the SRT executions is adapted by changing the frequency that SRT executions are started. Two different frequencies are used: one parameter governs the frequency of the nested object executions and another parameter governs the frequency of replicated SRT executions. The total number of relicated (Rtot) and nested SRT executions (Ntot) are presented for a total number of periods of 15.

In table 2 a few combinations of Rtot, Ntot and TOmlt are shown followed by the number of aborts (Nab), the number of retries caused by version overwriting (Nret) and the number of sucesses (Nsuc). The treatment of the request waits until the requesting CPU becomes gap manager. In table 3 the same combinations are shown, but now the SRT request for a SA executions is sent directly to the presently active gap manager. Both should be compared with table 1, which shows the same numbers when no concurrency control algorithm is executed. The few aborts are caused by overloaded CPUs.

In general, the total number of generated requests is not equal to the number of aborts, retries and successes: Nab+Nret+Nsuc $\neq$ Rtot+Ntot. At the end of the simulation (15 periods) not all requests have been terminated; additionally, one request can suffer more than one retry. It can be noted that the increase of TOmlt has little influence on the simulation results. This indicates that requests are either treated fast and correctly or are constantly prevented from a successful continuation. This is caused by the waiting for commit when a version is read.

The effect of changing the gap manager strategy demonstrated by the differences between tables 3 and 2, is more dramatic. Under the conditions of table 3 the treatment of commits is done more efficiently. Consequently, requests wait a shorter time before they are committed in a gap. Therefore, the read requests spend less time waiting for the commit of the versions. This reinforces the above conclusion about the causes for the high number of aborts.

In table 3 more retries (Nret higher) are detected. In this case the gap is used more efficiently by the SA executions effectuating the commits and less time is available for the SRT executions. Therefore, the chance of reading versions too late increases. From the simulation it was concluded that many aborts concern waiting read requests. The algorithm was modified such that read-only transactions are ordered with their start of execution timestamp with respect to the commit timestamp of the read-write transactions and not with respect to the start of the read-write transaction. Consequently, read-only transactions do not wait, but can proceed directly. These results are shown in table 4, where the modified algorithm is marked with read-no-wait rnw and the original algorithm with org for two cases: two and four versions per object. Read-only transactions are disadvantaged with respect to read/write transactions as the total life-time of the number of committed transactions is reduced. The difference between the two algorithms has vanished when four versions per object are used. To really profit from the modified rnw-algorithm, priorities need to be introduced, to render the read-only transactions completely independent of the read-write transactions.

6 Conclusions

ExSpect proves to be a useful tool to develop the concurrency control algorithm of DEDOS. Several strategies can be tried out before they are implemented, thus leading to significant development time reductions. The extension to the traditional VDM-like [12] specification with channels and timed tokens is well adapted to the specification of concurrency algorithms. Single PETRI-nets miss the possibility to include data-

Table 4: Number of successes (Nsuc), Aborts (Nab) and retries (Nret) as function of the parameters strategy (strat) and number of versions (Nver). Send request to gap manager strategy is used; Tomlt=20, Rtot=309, Ntot=306.

Strat	Nver	Nab	Nret	Nsuc
rnw	2	249	38	359
rnw	4	155	12	442
org	4	154	11	448
org	2	205	47	397

dependence in the specification, while the expression of concurrency is notoriously difficult to express in VDM and other similar languages. Languages like SDL [13] and LOTOS [14] allow the expression of concurrent processes as in EXSPECT, but they impose a certain type of message model. In the here presented example, EXSPECT is i.a. used to specify the message model.

However, it is important to note that the additional complexity of timed tokens makes the specification a challenging subject. Care is required during the specification to prevent race conditions.

The hierarchy of systems in ExSpect makes it possible to adapt small well isolated parts of DEDOS during the simulation phase before changing the actual system. For example, the allocation of a CPU to a SRT execution is on a first-come first-served basis. This scheduling can be changed (e.g. to shortest deadline) in a well isolated part (system) of the specification.

HRT transactions can continue and communicate with SRT transactions without perturbation by the SRT transactions. SRT read transactions can progress without any waiting time and their success rate depends on the load of the CPU's, the scheduling strategy and the number of versions. Concurrency control aborts only occur when two transactions have overlapping read and write sets. A careful construction of the application which prevents the simultaneous write into the same objects by two or more SRT executions assures that no waiting times are introduced by the concurrency algorithm. When two executions try to modify the settings of a piece of hardware simultaneously by modifying an object, then only one of them will succeed, while the other is aborted or retried. Just the kind of behavior required for real time systems.

References

[1] D.K. Hammer, E.J. Luit, O. van Roosmalen and P.D.V. van der Stok, " The Dependable Distributed Operating System DEDOS: an overview", *TUE computer Note*, (in preparation).

[2] J.C. Laprie, "Dependability: an unifying concept for reliable computing and fault- tolerance", T Anderson, (ed.), *Blackwell Scientific Publications*, Oxford, 1989.

[3] J.P.C. Verhoosel, E.J. Luit, D.K. Hammer, "A static scheduling algorithm for Distributed hard Real-Time systems", *The Journal of Real-Time systems*, Vol 3, pp. 227- 246, 1991.

[4] B. Meyer, "From structured programming to object-oriented design: The road to Eiffel", *Structured programming*, Springer Verlag, 1989.

[5] P.D.V. van der Stok and A. Engel, "Shared data concepts for DEDOS", *Proc. of the 10th IFAC workshop of Dist. Comp. Control Systems*, Semmering, Austria, 9-11 Sept. 1991, IFAC Workshop series, Ed. H.Kopetz and M.G.Rodd, No 3, Pergamon Press, 1992.

[6] P.D.V. van der Stok, O.S. van Roosmalen, E.J. Luit and D.K. Hammer, "An object-oriented approach to edependable responsive systems", *1st Int. Workshop on Responsive Systems*, Golfe-Juan, France, 3-4 Oct. 1991.

[7] D.P. Reed, "Implementing atomic actions on decentralized data", *ACM Transactions on Computer Systems*, Vol 1, pp 3-23, 1983.

[8] K.M.van Hee, L.J.Somers, M.Voorhoeve, "Executable specifications for distributed information systems", E.D.Falkenberg, P.Lindgreen (eds.), *Information system concepts: an in-depth analysis*, North-Holland, 1989.

[9] K.Jensen, "Colored Petri Nets: A High Level Language for System Design and Analysis", G.Rozenberg (ed.), *Advances in Petri Nets 1990*, Lecture Notes in Computer Science 483, Springer Verlag, 1991.

[10] F. Cristian, "Understanding fault-tolerant distributed systems", *Communications of the ACM*, Vol 34, no 2, pp 57-78, 1991.

[11] P.T.A. Thijssen, "Specification of DEDOS Concurrency Control with ExSpect", *EUT master's thesis*, Dec. 1991.

[12] C.B. Jones, "Systematic software development using VDM", Prentice-Hall, 1986.

[13] CCITT, Recommendations Z.100-Z.104, Vol VI, Fasc VI.10, 1984.

[14] ISO, DP8807rev, "LOTOS: A formal descrition tecnique based on temporal ordering of observational behavior", Geneva, 1988.

Panel Discussion

What Are the Key Paradigms in the Integration of Timeliness and Availability ?

F. Cristian

Computer Science and Engineering ,
University of California, San Diego, USA
619-534-6383, flaviu@cs.ucsd.edu

We address this issue in the context of distributed responsive systems. Their key goal is to provide users with services that are *responsive*, that is, are *timely* and *fault-tolerant* under specified load and failure hypotheses. A user might in its turn provide services to other users, so there can be a complex *depends-upon* relation between services situated at different levels of abstraction.

To make a service responsive to a user, it is sufficient to ensure that the *communication* service between them is responsive and that the *service* itself is responsive.

To ensure that a service is fault-tolerant a key paradigm is to implement it by a *group of* redundant, independent *servers*. To ensure that a redundant service implementation is timely, one must first establish the timeliness of individual servers. This means establishing that each server reacts to service requests subject to certain *deadline* and *jitter* constraints. Little is known about how to systematically derive such constraints for an arbitrary service at an arbitrary level of abstraction from top-level specifications. However, once such constraints are formulated, one can use scheduling theory results to ensure that a server satisfies them. The main paradigms available for this are *off-line, static* scheduling and *on-line, dynamic* scheduling.

The two approaches that dominate static scheduling are the *cyclic executive* paradigm, in which a fixed set of tasks is executed cyclically at a fixed rate, and the *rate monotonic scheduler* paradigm, in which tasks are assigned priorities proportional to their rates (or frequencies) and the ready task with the highest priority is chosen for execution at any time. There is an ongoing debate concerning which of these two approaches is better. Among the drawbacks of the cyclic executive (or *static scheduling*) approach, the most frequently cited are 1) wastefulness of resources (sufficient time must be allocated statically to satisfy timeliness under worst case load and worst case failure assumptions), 2) fragility when lower timing assumptions are violated (continuing the computation will delay all following activities, aborting it can create inconsistent states and cause mishaps in critical systems), and 3) high costs associated with change and maintenance (the timing analysis has to be re-done when the arrival rate of an event changes or some new event class must be accommodated). The key advantages claimed by the proponents of the cyclic executive approach are 1) its determinism under stated worst case load and failure hypotheses and 2) its conceptual simplicity. Proponents of the rate monotonic approach claim the following advantages: 1) its capacity for graceful degradation when timing assumptions are violated and 2) its maintainability when new events must be taken into account or event frequencies change. Among the criticism made to rate monotonic scheduling, the most important is the persistence of certain restrictive hypotheses affecting the applicability of the theory. For example, no widely accepted results exist for scheduling tasks on multi-processors with true concurrency or in environments in which the underlying hardware, operating and communication systems can fail.

In contrast to static scheduling, where a priori knowledge of task deadlines and their maximum frequency must be available before execution, dynamic schedulers need not know anything about the tasks to be scheduled until they arrive. When they enter the system, their deadlines are used explicitly by the (on-line) scheduler to generate -if possible- a schedule. When a schedule is not feasible for a set of tasks, the goal will be to minimize the damage caused by some tasks missing their deadline. Among the on-line schedulers, the most prominent is the *earliest-deadline-first* algorithm, known to be optimal for the case of an underloaded uniprocessor: if any scheduler can meet all the deadlines of a given set of tasks, then the earliest-deadline-first can too. It is unfortunately known that the earliest-deadline-first (as well as the smallest slack time first) algorithms perform quite poorly when the system is overloaded. Under overload conditions, it has been proven that no on-line scheduling algorithm will achieve a competitive factor of more than 0.25 (where the competitive factor measures how good the on-line algorithm is compared to a clairvoyant algorithm that knows everything about the future). Algorithms that achieve the best theoretically possible competitive factors for independent tasks have been proposed recently, but due to the different assumptions they make, it is hard to characterize how these algorithms perform under overload compared to a rate monotonic algorithm that say deals with task synchronization and bounds priority inversions. Such a comparison should add new momentum to the ongoing debate between defendants of static scheduling and proponents of dynamic scheduling, debate mostly caused by the different assumptions made by these two approaches.

Once the timeliness of the service provided by a single server is established, one can proceed to establish the timeliness of the service provided by a group of servers. Depending on whether the implementation of the service is loosely synchronized or tightly synchronized, the problem can be more or less simple. In a *loosely synchronized* implementation, one server, the *primary*, maintains the current service state while the others (the *backups*) maintain past states and log the service requests addressed to the primary. To maintain logs finite, the primary periodically checkpoints its state to the backups, who then discard their logs. In the *closely synchronized* case, all servers interpret all service requests in parallel, so that their local states are close to each other. The simplest case for establishing timeliness of a server group is when the members are closely synchronized and they all respond to each service request: if individually the servers are timely, they will be timely as a group. The main drawback of this approach is waste in communication bandwidth. To economize bandwidth, group members can be ranked with respect to communication so that only the highest ranking sends replies. This introduces a substantial complication: one has to establish that a primary failure is detected by a backup in a timely manner, so that it can continue to send replies to users. The group communication service that ensures detection and agreement upon member failures is termed the *group membership* service. Group membership is also needed for loosely synchronized servers, where a backup must detect a primary failure in a timely manner to be able to begin interpreting its log of requests, reach a state consistent with the state of the primary before the crash , and continue to send replies from that point on. Assuming the timeliness of the failure detection and log consumption established under worst case load and failure assumptions, the next problem is to ensure that there are *enough* redundant servers for each service in a system, so that the likelihood of all servers failing simultaneously is negligible with respect to the stated service availability requirements. The general service that automatically maintains a quasi-constant population of servers for all services provided by a system is termed a *service availability management* service. This service must itself be timely and available, to ensure prompt reaction to any failures and joins. This can be achieved by replicating global system state information on several of the processors of a system. The general group communication service used to broadcast global state updates to a server group and maintain the consistency of their local states is termed *atomic broadcast*.

To ensure that communication between users and servers is available, the two main implementation paradigms are *acknowledgement* based protocols and *diffusion* based protocols. Many practical protocols are in fact combinations of these approaches. In the acknowledgement based approach, time redundancy is used to re-send an information until the sender learns (from a server membership service) that the servers for which the information was intended have received it, have failed or are partitioned from the sender. Acknowledgement based communication can be further classified into *positive acknowledgement* based and *negative acknowledgement* based. While positive acknowledgements provide in general less jitter in message delivery delay at the expense of an increased message traffic in the absence of failures or overload, negative acknowledgements save messages when there is no overload or failures at the expense of providing more jitter when messages are not frequent enough. In the diffusion based approach, information is sent via several independent routes in parallel, so that at least a message carrying it arrives at each destination under the assumed failure and load hypotheses. Because of lack of message re-transmission, the diffusion based communication has little or no jitter. One drawback of the diffusion based approach is that it sends more messages than an acknowledgement based approach when everything works well. On the other hand, an acknowledgement based communication sends more messages when delay hypotheses are exceeded and when the system is overloaded, thereby creating more overload and longer delays, which can lead to further overload and longer delays, all leading to eventual congestion and partition. There exist both diffusion based and acknowledgement based implementations of group services such as the atomic broadcast, membership and service availability management services mentioned before, but little is known about how these implementations compare in practice. To specify the routes on which messages travel, the two main paradigms are *static routing* and *dynamic routing*. While static routing is simple to understand and makes the determination of timeliness properties easier, it lacks flexibility and can be wasteful of bandwidth. Dynamic routing ensures a better utilization of the available communication bandwidth and adapts better to failures, but exacerbates the difficulty of establishing timeliness and doing flow control.

The subject of establishing the timeliness of information arrival at a destination in the presence of competition for communication resources and communication failures is less well understood than the problem of checking the timeliness of a classical, non-distributed service on a uniprocessor. In principle, one can adopt either a *static* approach, such as Time Division Multiplexing, or a *dynamic, on-line* approach based on message deadlines, with drawbacks and advantages similar to the ones known for the static and dynamic approaches mentioned earlier. As in the case of classic scheduling, static approaches might seem more natural for a fully predictable environment, where all events to be handled are known apriori, are cyclic, and occur with fixed frequencies and very little jitter, while dynamic approaches might be preferred when the number of events to handle is huge and load is not so predictable. The problem of establishing the timeliness of local and long-haul communication services has recently acquired an increased importance because of the availability of multimedia applications. Much of the research performed these days on timely communication is done within this community.

Contribution to the Panel: What are the Key Paradigms in the Integration of Timeliness and Availability ?

G. LE LANN

INRIA,France

Availability

Availability implies redundancy. Redundancy management raises synchronization issues.

One approach, the Synchronous approach, rests on the assumption that some underlying subsystem provides for some (possibly elementary) global synchronization, e.g. a time-slotted bus.

The other approach, the Asynchronous approach, precisely addresses the problem of how to bring asynchronous entities (processors, processes, etc.) into physical/logical synchrony.

It is our view that the transition from asynchronism to synchronism raises very fundamental issues. Algorithms used to accomplish this transition fully determine overall system behavior (they could be viewed as system "genetic codes"). For example, would centralized polling or centralized clocking be used by the underlying subsystem, the overall system inevitably belongs to the category of centralized systems. Some systems, called distributed, which resort to "centralization hiding" via additional hardware/software layers, are in fact inherently centralized.

One of the major drawbacks of centralized synchronization algorithms is that they are not fault-tolerant. Whenever it is attempted to make them fault- tolerant (e.g. primary/backup schemes), one is faced with the general synchronization problem again! General solutions are known, namely decentralized synchronization algorithms.

Hence, the conclusions:

- redundant architectures (hardware, software, data), which are needed
  to obtain availability, raise decentralized synchronization issues similar to those raised by distributed systems,

- such issues are addressed under the Asynchronous approach,

- such issues are assumed to be addressed "elsewhere" under the Synchronous approach,

- from an algorithmic viewpoint, redundant systems designed according
  to a Synchronous approach either are incorrect (inadequate solutions
  come with the underlying subsystem) or are correct but boil down
  to tautologies (higher synchronization is obtained out of preexisting
  lower synchronization).

Therefore, it must be clear that the design and building of available
(i.e. redundant) systems cannot be conducted correctly without selecting or defining – and using – at least one decentralized synchronization
algorithm.

Timeliness

The basic issue raised with timeliness is whether or not it is possible to
know A PRIORI that:

P1 : all task timing constraints WILL be satisfied, if the scheduling
problem considered has an exact solution (the task set considered is
schedulable),

P2 : a value/reward function WILL be optimized/maximized if the
scheduling problem considered has no exact solution (e.g., the least
critical tasks are abandoned when considering a non schedulable
task set).

Let us call predictability (P) this property of a priori knowledge [P=P1
U P2].
There are three essential concepts involved with this property, namely
"optimality", "distance from optimality" and "schedulability analysis".

To prove/check that P holds, in addition to describing the scheduling algorithm(s) S considered, one needs to define the following three models:

T : task model (resource requirements, execution time, deadline, etc.),

A : arrival law model (periodic, sporadic, aperiodic arrivals),

R : resource model (processors, storage, network, data structures, de- vices, etc.).

It may be that a unique scheduling algorithm is used to allocate resources to tasks, or that different algorithms are used for allocating different resources.
A good design in (timeliness) predictability is such that the following implication holds :

$$S*T*A*R => P$$

Approaches to this design problem vary greatly. Since the early 70's, where simple models were considered, more complexity has been analyzed and mastered.
It appears that the most controversial issues currently are :

- off-line versus on-line scheduling

- is P2 an interesting property ?

We argue that with distributed/replicated systems using decentralized control (for the sake of availability), such controversies are meaningless. We argue that, from a theoretical viewpoint, there is no choice left:

- yes, on-line scheduling is needed,

- yes, P2 is an interesting property.

First, we can demonstrate that off-line S algorithms rest on T, A and R models which are too simple, and antagonistic with distributed systems. For example, it is assumed that exact task execution times are known a priori, that tasks are independent. When resource sharing is considered, exact access times and utilization times are assumed to be known a priori for all tasks. Or only periodic tasks are assumed to be bound to meet strict timing constraints.

Second, it is well known that beyond these simple T, A and R models, no optimal polynomial off-line scheduler exists. Consequently, the optimality criterion cannot be used to reject on-line schedulers.

Third, simple T and R models do not allow to cope easily with failures. For example, task execution times and task/processor mapping may become variable rather than constant. When execution times and mappings vary, it becomes intractable to predict all future timing patterns for external and internal events.

Fourth, simple A models may lead to fallacious predictability. Simple A models, e.g. periodic arrivals "resulting from" the decision to periodically sample the environment, are equivalent to ruling out the existence of future "overload" situations. What is the validity of such "predictions"? Systems designed under this kind of approach "break down" whenever faced with a supposedly impossible situation. Conversely, realistic (i.e. complex) A models very often lead to adopting on-line schedulers. Contrary to what happens when using off-line schedulers and simple A models, on-line scheduling based systems never face "impossible" situations.

Fifth, on-line schedulers are generally characterized by their "distance" from optimal schedulers. This is how, in particular, P2 is ensured. For example, such a "distance" has been shown to have an upper bound in the case of aperiodic tasks, characterized by a deadline and a value. Such models represent problems that are beyond the capabilities of off-line schedulers. Consequently, only on-line scheduling can help characterize predictability in realistic cases.

A last note is in order regarding schedulability analysis. It is sometimes argued that predictability cannot be obtained with on-line schedulers, for the reason that schedulability analysis is feasible only with off-line schedulers.

This is an erroneous view, which confuses conclusions and assumptions. On the contrary, it is possible indeed to conduct a schedulability analysis with an on-line scheduler, under T, A and R models identical to those considered with off-line schedulers. Predictability can be obtained with on-line schedulers for and beyond these models.

Complicated Paradigm of Responsive Systems

Y. TOHMA

Department of Computer Science
Tokyo Institute of Technology, Japan
E-mail:tohma@cs.titech.ac.jp

Questions related to responsive systems will be argued. This argument hopefully help us deepen the insight where problems reside and how these are difficult.

1 Environment

Since the responsive system necessarily handles the relationship between the applied stimulus and the generated response, it cannot exist alone. The *communication* is essential to responsive systems.

The communication will be made in different ways. The simplest one is the one-to-one communication, while the group(n)-to-group(m) communication may also take place. Moreover, not only such pairwise communications but also multi-group/individual communication encounters simultaneously. In the latter case, the difficult problem such as maintaining the *consistency* among the members of the group may arise.

2 Behavior

In modern engineering world, the interpretation of the response of a system is not straightforward. The response will be recognized by *events*. Is the response constituted by a single event or a sequence of events? Which is to be considered obviously relates to the nature of events. Further, when considering the sequence of events, is the occurrence of events *synchronous* or *asynchronous*, otherwise *clocked* or *self-timed*?

When asynchronous and/or self-timed operation is considered, the provision and the satisfaction of timing constraints will be very complicated issues.

The dimension of synchronism vs asynchronism should be distinguished from that of clocked vs self-timed operation. We should discuss in more detail what synchronous (or asynchronous) systems are.

3 Attributes

An event may have several attributes such as *value, time,* and *dependability measure,* even when the nature of events is defined a priori. Namely, the event may be denoted with its attributes as

$$E(\mathbf{v}, \mathbf{t}, \mathbf{f})$$

where $\mathbf{v}$, $\mathbf{t}$, and $\mathbf{f}$ represent the value, the time , and, say, the failure probability, respectively. These variables may be scalars or vectors.

The major concern of responsive systems is obviously the *timing* relationship with counterparts. However, all members involved in the communication may not necessarily be synchronized with the global (reference) time. Most responsive systems will care only the time duration elapsed between the stimulus and the response. When events occur in terms of the global time may be out of interest in many cases.

When a coordinated operation should be initiated autonomously (without being caused by a conductor) by each member of a group at the same time, the *synchronization* of clocks of each member of the group is, of course, required.

4 Description and Analysis

How should the event of a responsive system be described in terms of its attributes, particularly with the timing constraint? What models are available? The finite-state machine model can be extended to include the timing constraint. The timed Petri net have been proposed. Are all of these models effective in describing real responsive systems?

When possible sequences of events of multiple responsive systems mutually interacted is analyzed, we may readily encounter the enormous number of states to be examined. It looks a difficult but challenging issue to find ways to alleviate the so-called *state explosion* in such analyses.

5 Failure Mode

The failure probability as an attribute parameter of the event needs more detailed consideration. In a way, it can concern the pair of **v** and **t**. It represents the probability that the correct occurrence of the both of **v** and **t** fails. However, **f** may be split into components which relate to **v** and **t**, respectively. The difficulty here is that the failures at **v** and **t** may correlate or may be mutually independent.

Since the timeliness is key issue of responsive systems, we should note more timing failure such as missing of the response, generation of spurious response without any cause, the occurrence of response at an improper time, etc.

6 Fault Tolerance

For the detection of timing failure, we need the *reference of time* as a basis of the watchdog timer. In many cases, however, how to evaluate the proper reference of time is not so simple matter.

Can the technique of retry be applied to tolerate the timing failure? In order to meet rigorous constraint of time in the response, check points should be inserted frequently so that the performance of functionality may degrades.

It looks very hard to correct the timing failure in contrast to the case of the error correction of response value. Robust timing window and the redundant constituent units would be needed not to provide the counterparts with improper response, but forward only the response at a proper time from healthy unit. More new approach to the *timing fault-tolerance* would be required.

Toward Responsive Distributed Systems

M. Tokoro

Sony Computer Science Labolatory, Japan

Abstract

Computing environments in the next decade will be widely distributed, ever-changing, and ubiquitous. People will move with computers and will use them while moving. People will demand better user interfaces, so that they will be able to communicate with computers as if they are communicating with other people. For such purpose, we propose a hand-held computer called an Intimate Computer. It is a personified computer virtualizing the sincere spouse, reliable secretary, and/or dependable buddy of each individual. People will carry it, talk to it, and it will reply with a smile. It works an access terminal to a widely distributed computational resources. For such mobile computers to be materialized, we propose the Computational Field Model, which is a higher level abstraction of distributed computing systems. It is based on the notion of concurrent objects and provides an asynchronous and real-time computing environment.

In such a widely distributed computational environment, the notion of Responsive Systems shows different views from that of centralized systems. It is an open system which are shared by many people. Therefore, anything can happen in such an environment. We cannot premise any "predictability in the system. We have try our best not to say that "it is out of the specification, crash". The real time feature is necessary because the system is connected with our everyday life. So as the feature of fault tolerance.

We propose an object-oriented framework for Distributed Responsive Systems. It is based on the notion of "Best Effort and Least Suffering". An object which receives a request does its best effort to satisfy the sender. On the other hand, an object which sends a request does its best to survive, or, in the other words, not to hang up. This is summarized as that an object as a server does its best effort while an object as a client does its least suffering. Since an object in a distributed system acts as a server and a client at the same time, each object has to provide the property of Best Effort and Least Suffering.

We have developed a real-time network protocol called RtP and a real-time programming language to describe distributed systems called DROL based on the above mentioned notion. These attempts are our initial step toward true Distributed Responsive Systems, so that we can have responsive intimate computers in the responsive computational field.

Responsive System Design

A Reconfigurable Parallel Processor Based on a TDLCA Model

M. Tsunoyama[a], M. Kawanaka[b], S. Naito[c]

a Nagaoka College of Technology

b NEC Robotics Engineering Ltd.

c Tokyo Metropolytan University

Abstract

This paper proposes a reconfigurable parallel processor based on two- dimensional linear cellular automaton model. The processor based on the model can be reconfigured quickly by utilizing the chharacteristics of the automaton used for its model. Moreover, the processor has short data path length between processing elements compared with the length of the processor based on one-dimensional linear cellualr automaton model which has been already discussed.

The processing elements of the processor based on the two-dimensional linear cellular automaton model are regarded as cells and the operational states of the processor are treated as the states of the automaton. When faults are detected, the processor can be reconfigured by changing its state under the state transition function of the processor determined by the weighting function of the automaton model. The processor can be reconfigured within a clock period required for making a state transition. This processor is extremely effective for real-time data processing systems required high reliability.

1   Introduction

Several types of highly parallel processors, such as array type and tree type, have been proposed and some of them are already used for high speed data processing systems[1, 2]. These parallel processors contain many processing elements, so that the probability of faults in the processor may increase and they can not preserve their high availability.

To guarantee the high availability, they must be reconfigurable for replacing the faulty processing elements to normal ones when faults are detected. Several reconfiguration schemes have been proposed for highly parallel processors, however, they require rather long time to complete the reconfiguration since they employ complicated reconfiguration algorithms[3, 4]. When the parallel processors are used in a highly reliable real time data processing systems, the processor must be reconfigured quickly, otherwise the input data during the reconfiguration will be lost and the system can not output the correct result.

This paper proposes a reconfigurable parallel processor based on a two dimensional linear cellular automaton(TDLCA) model. In the model, the processing elements of the processor are regarded as the cells of the automaton and the operational states of the processor are treated as the states of the automaton. When faults are detected, the states of the faulty processing elements are changed to the states for non-operating processing elements under the state transition function of the processor. The state transition function is made by the weighting function of the TDLCA used for its model. After the state transition, the faulty processing elements are replaced logically by normal ones and normal elements are reconnected by using labels determined by states of neighboring processing elements. The processor can immediately restart the data processing after the state transition.

In this paper, section 2 shows the definition of a TDLCA and some properties of it. The states of the cellular automaton satisfying conditions shown in the section are M-planes and they have constant-weight and equidistance(CWED) properties. Section 3 shows the definition and some concepts of the parallel processor based on a TDLCA model. The parallel processor proposed in this paper consists of the two major blocks: global control block and processing element block. Each processing element in the processing element block consists of a data processing part and a local control part. The local control part controls the data stream using the labels of the processing elements and the relative addresses from the source processing element to the destination ones. In this section, the reliability is also consided. This reconfigurable parallel processor can preserve higher reliability than the duplex parallel processors which uses about the same amount of hardware as the proposed processor. Moreover, the processor has shorter data path length between processing elements and can process input data faster than the processor based on one-dimensional linear cellular automaton model already proposed[5, 6]. The final section 4 is the conclusion.

2 TDLCA model

2.1 Basic Definitions of TDLCA

A two dimensional cellular automaton consists of many identical automaton called cells. These cells are placed on a plane and the automaton is defined as follows.

Definition 1

$$M_2 = (Z_{n_1,n_2}, Q_2, W_2) \tag{1}$$
$$Z_{n_1,n_2} = Z_{n_1} \times Z_{n_2}$$
$$Z_{n_1} = \{0, 1 \cdots n_1 - 1\}, Z_{n_2} = \{0, 1 \cdots n_2 - 1\}$$
$$Q_2 = \{q_0, q_1 \cdots q_{t_2-1}\}$$
$$W_2 = S_{s_2} \to S_{s_2}$$

$$S_{s_2} = \left\{ S_2 = \begin{bmatrix} s_{0,0} & \cdots & s_{0,n_2-1} \\ s_{1,0} & \cdots & s_{1,n_2-1} \\ & \cdots & \\ s_{n_1-1,0} & \cdots & s_{n_1-1,n_2-1} \end{bmatrix} \right\}$$

In the above definition, the set Z_{n_1,n_2} is a direct product of Z_{n_1} and Z_{n_2}. These two sets are the residue classes of an integer ring modulo n_1 and n_2, respectively. The set Z_{n_1,n_2} is a set of ordered pairs which represents the positions of the cells. The set Q_2 is a finite set of states of each cell and a finite field GF(q) is used to represent these states in this paper where q is a prime number. Set S_{s_2} is a finite set of states of the automaton and each element of the state(viz. $s_{0,0}, s_{0,1} \cdots$) is a state of each cell. The state is represented by a matrix. W_2 is a mapping from the set of states S_{s_2} to itself and is called an weighting function of the automaton. The state of every cell in the automaton changes synchronously under the weighting function W_2. A two dimensional cellular automaton is called a two-dimensional linear cellular automaton and denoted by TDLCA when its weighting function is linear as shown by the following definition.

Definition 2

$$s_{i_1,i_2}\prime = \sum_{j_1+k_1=i_1} \sum_{j_2+k_2=i_2} w_{j_1,j_2} \cdot s_{k_1,k_2} \tag{2}$$

$$0 \le i_1, j_1, k_1 \le n_1 - 1$$
$$0 \le i_2, j_2, k_2 \le n_2 - 1$$
$$[s_{i_1,i_2}\prime] \in S_2\prime$$
$$w_{j_1,j_2}, s_{k_1,k_2} \in Q_2$$

From Eq.(2), we can say that the next state $S_2\prime$ is obtained by a con-volution between W_2 and the current state S_2. In this case, the next state can be calculated by a product of two variable polynomials when an weighting function W_2, current state S_2 and the next state $S_2\prime$ are represented by two variable polynomials[7]. An weighting function W_2 and a state S_2 in the forms of two variable polynomials are shown by Eq.(3).

$$W_2(x, y) = \sum_{i=0} \sum_{j=0} w_{i,j} \cdot x^i \cdot y^j \tag{3}$$

$$S_2(x, y) = \sum_{i=0} \sum_{j=0} s_{i,j} \cdot x^i \cdot y^j$$

The next state $S_2\prime(x, y)$ represented by the two-variable polynomial can be obtained by the product of the above polynomials $W_2(x, y)$ and $S_2(x, y)$. These polynomials are the members of the residue class of the two variable polynomial ring of modulo $(x^{n_1} - 1, y^{n_2} - 1)$ since the Z_{n_1,n_2} is a direct product of two residue classes of an integer ring Z_{n_1} and Z_{n_2}, respectively. In this paper, we use the notation W_2 and S_2 for the polynomial representation of $W_2(x, y)$ and $S_2(x, y)$.

2.2   TDLCA Having CWED Properties

Two kinds of distances between two states $S_2^{(i)}$ and $S_2^{(j)}$ ($S_2^{(i)}, S_2^{(j)} \in S_{s_2}$) of the TDLCA $d_H(S_2^{(i)}, S_2^{(j)})$, and $d_L(S_2^{(i)}, S_2^{(j)})$ are defined as follows:

Definition 3

$$d_H(S_2^{(i)}, S_2^{(j)}) = \sum_{i=0} \sum_{j=0} d_{h_{i,j}} \tag{4}$$

$$d_{h_{i,j}} = \begin{cases} 1 & S_{2_{k,l}}^{(i)} \neq S_{2_{k,l}}^{(j)} \\ 0 & S_{2_{k,l}}^{(i)} = S_{2_{k,l}}^{(i)} \end{cases}$$

$$0 \leq k \leq n_1, 0 \leq l \leq n_2$$

$$d_L(S_2^{(i)}, S_2^{(j)}) = \sum_{i=0} \sum_{j=0} d_{l_{i,j}} \tag{5}$$

$$d_{l_{i,j}} = \begin{cases} S_{2_{k,l}}^{(i)} - S_{2_{k,l}}^{(j)} & S_{2_{k,l}}^{(i)} \geq S_{2_{k,l}}^{(j)} \\ S_{2_{k,l}}^{(j)} - S_{2_{k,l}}^{(i)} & S_{2_{k,l}}^{(i)} \leq S_{2_{k,l}}^{(j)} \end{cases}$$

The weight of a state is defined as the distance d_H between the state and the zero state in which all elements are zero. A TDLCA is called TDLCA with the constant-weight and equidistance(CWED) properties if the weight of all states are constant and the distances d_H (and d_L) between any two states of them are identical.

One dimensional linear cellular automaton LCA ($M = (Z_n, Q, W)$) can be isomorphic to a TDLCA $M_2(= (Z_{n_1,n_2}, Q_2, W_2))$ and is obtained by an isomorphism ϕ shown by the following equation[8, 9].

$$\phi(M_2) = (\phi_Z(Z_{n_1,n_2}), \phi_S(Q_2), \phi_W(W_2)) = M \tag{6}$$

where

$$\phi_Z((c_1, c_2)) = c_1 \cdot n_2 \cdot t_1 + c_2 \cdot n_1 \cdot t_2 \bmod (n_1 \cdot n_2),$$

$$(c_1, c_2) \in Z_{n_1,n_2},$$

$$n_2 \cdot t_1 = 1 \bmod n_1$$
$$n_1 \cdot t_2 = 1 \bmod n_2$$

n_1 and n_2 are relatively prime and

$$n_1 \cdot n_2 = n = q^t, t \in Z$$

In Eq.(6), ϕ_S is an identity mapping from Q_2 to Q, and ϕ_W is made by the mapping ϕ_Z as follows:

$$\phi_S(Q_2) = Q$$

$$\phi_W(W_2) = \sum_{i=0}^{n_1-1} \sum_{j=0}^{n_2-1} w_{\phi_Z(i,j)} \cdot x^{\phi_Z(i,j)}$$

Therefore, a TDLCA satisfying the condition shown in the following theorem can have the CWED properties[10, 11].

Theorem 1 *TDLCA*

$$M_2(= (Z_{n_1,n_2}, Q_2, W_2))$$

has CWED properties, if $H(x) = \frac{(x^{n_1 \cdot n_2} - 1)}{W(x)}$ *is a primitive polynomial,* $W(x) - x^t \in < H(x) >, t \in Z$, *and the integers,* t *,n_1, and n_2 are relatively prime.*

Example 6 *TDLCA* $M_2 = (Z_{5,3}, GF(2), W_2)$

$$W_2 = 1 + x \cdot y + x^2 \cdot y^2 + x^3 + y^2 + x^2 \cdot y + x^2 \cdot y^2 + x \cdot y^2$$

has the CWED properties and states of it are given by

$$S_2^{(0)} = \begin{bmatrix} 1 & 0 & 0 & 1 & 0 \\ 0 & 1 & 1 & 0 & 0 \\ 1 & 1 & 1 & 1 & 0 \end{bmatrix}$$

$$S_2^{(1)} = \begin{bmatrix} 1 & 1 & 0 & 0 & 0 \\ 1 & 1 & 1 & 0 & 1 \\ 0 & 0 & 1 & 0 & 1 \end{bmatrix}$$

$$\cdots$$

$$S_2^{(14)} = \begin{bmatrix} 0 & 1 & 1 & 1 & 1 \\ 0 & 1 & 0 & 0 & 1 \\ 0 & 0 & 1 & 1 & 0 \end{bmatrix}$$

The LCA M being isomorphic to TDLCA M_2 is obtained by the isomorphism ϕ shown by Eq.(6) where $n_1 = 5$, $n_2 = 3$, $t_1 = 2$ and $t_2 = 2$.

$$
\begin{aligned}
M &= (Z_{15}, GF(2), W) \\
Z_{15} &= \{0, \cdots, 14\} \\
W &= \phi_W(W_2) = 1 + x + x^2 + x^3 + x^5 + x^7 + x^8 + x^{11}
\end{aligned}
$$

$$
\begin{aligned}
S^{(1)} &= (111101011001000) \\
S^{(2)} &= (111010110010001)
\end{aligned}
$$

$$\cdots$$

$$S^{(14)} = (011110101100100)$$

3 Reconfigurable Parallel Processor

3.1 Two Dimensional Parallel Processor

A reconfigurable parallel processor, RP_{s_1,s_2}, proposed in this paper is defined as follows.

Definition 4

$$RP_{s_1,s_2} = (P_{s_1,s_2}, Q_p, D, I, R) \tag{7}$$
$$P_{s_1,s_2} = \{p_{i,j}\}$$
$$Q_p = \{q_0, q_1 \cdots q_{r_2-1}\}$$
$$D = S_p \to S_p$$
$$I \subset Q_p$$
$$R : PID \to CNT$$

$$0 \le i \le s_1 - 1, 0 \le j \le s_2 - 1$$

$$S_{p_s} = \left\{ S_p = \begin{bmatrix} s_{p_{0,0}} & \cdots & s_{p_{0,s_2-1}} \\ s_{p_{1,0}} & \cdots & s_{p_{1,s_2-1}} \\ & \cdots & \\ s_{p_{s_1-1,0}} & \cdots & s_{p_{s_1-1,s_2-1}} \end{bmatrix} \right\}$$

In the definition, P_{s_1,s_2} is a set of $s_1 \cdot s_2$ processing elements and Q_p is a finite set of states for processing elements. Mapping D determines the state transition of the processor. S_{p_s} in this definition is a set of states of the processors and is represented by set of matrices. Each element of a state matrix is a state of a processing element. Symbol I represents a set of states for non-operating processing elements and is a subset of Q_p. Mapping R determines the processing elements to which the data is transmitted. We call it a routing R and can be changed according to the algorithm to be used in the processor. PID is a set of labels of processing elements and CNT is a set of relative addresses from the source processing element to the destination ones. Every active processing element has an unique label determined by the neighboring processing elements as shown in the following section. The set of relative addresses CNT consists of ordered pairs $(r, s)(0 \leq r \leq s_1 - 1, 0 \leq s \leq s_2 - 1)$ which represents the relative address of destination cells. When a processor $RP_{s_1,s_2}(= (P_{s_1,s_2}, Q_p, D, I, R))$ and a TDLCA $M_2(= (Z_{n_1,n_2}, Q_2, W_2))$ satisfy the following conditions, the RP_{s_1,s_2} can be constructed by using the M_2 as its model[11],[12].

Condition 1   *1. The number of processing elements in the processor is equal to the number of cells of the TDLCA M_2.*

   2. There exists a bijection mapping from the set Q_p to the set Q_2.

   3. The state transition function D is isomorphic to the weighting function W_2.

A reconfigurable parallel processor constructed by using TDLCA consists of two major blocks: global control block and processing element block. The global control block holds the operational states of every processing element. The processing element block contains n processing elements and each processing element consists of a data processing part and a local control part as shown in Fig.1. This local control part controls the data transmission using the labels of processing elements and the relative addresses from source prcessing elements to the destination ones.

3.2   Fault Model

Faults in the reconfigurable parallel processor is assumed to be:

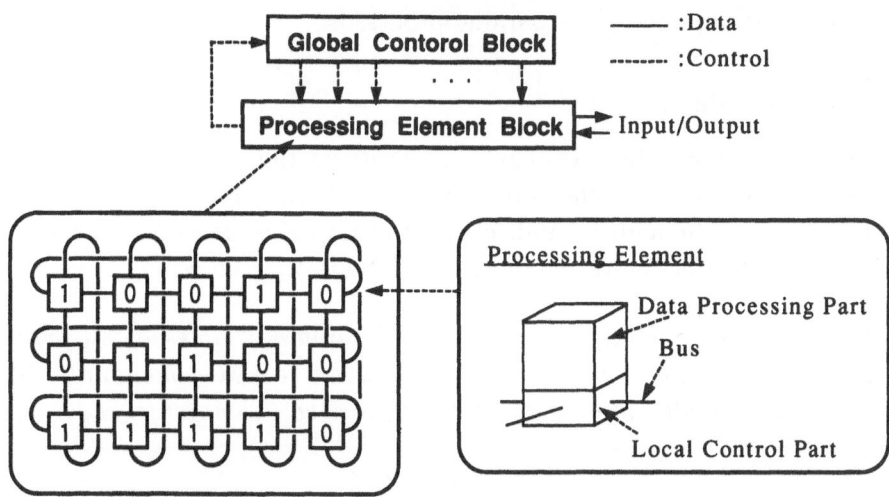

Figure 1: Block Diagram.

Assumption 1 *1. Each processing element i.e. a data processing part and a local control part, can detect its own faults.*

2. In the case of q=2, the maximum number of faulty processing elements in a processor is $log_2(n + 1) - 1$, where n is the number of processing elements in the processor.

3. Faults in the global control block can be detected.

When self-checking circuits are used for the processing element, assumption (1) can be satisfied. The number of cells at the state of an element of GF(2) which can be assigned to the states of faulty processing elements at the arbitrary position is limitted to $log_2(n + 1) - 1$ since the mapping ϕ in Eq.(6) is isomorphic and states of LCA form M-sequences[12]. The state matrices of TDLCA form a subset of two dimensional cyclic codes[14]. Faults in the global control block can be detected by using the technique developed for the two dimensional error detecting and/or correcting codes[15] since the state transition function of the processor is isomorphic to the weighting function of the TDLCA. Therefore, the assumptions from (1) to (3) are considered to be reasonable for implementing the reconfigurable parallel processor.

3.3   Reconfiguration

The processor can be reconfigured by setting the states of the faulty processing elements to the states in the set I when faults are detected. In this paper, the additive unit element of GF(q) is treated as the set I. The definition for the reconfiguration of an RP_{s_1,s_2} follows.

Definition 5 *An RP_{s_1,s_2} is called reconfigurable if states of all faulty processing elements can be assigned to the additive unit element of GF(q) and the non-fualty processing elements can be reconnected under the routing R.*

For connecting non-faulty processing elements to execute a specified algorithm, every normal element must have unique labels. In RP_{s_1,s_2} every processing element can have unique label based on the characteristics of the state of the processor since following theorem holds[16].

Theorem 2 *The states for the reconfigurable parallel processor based on a TDLCA model satisfying the condition in Theorem 1 are M-planes.*

From the above theorem, a label for a processing element can be determined by the following manner using the states of neighboring processing elements. A label $id_{i,j}$ for a processing element $p_{i,j}(0 \le i \le s_1 - 1, 0 \le j \le s_2 - 1)$ is given by a matrix

$$id_{i,j} = \begin{bmatrix} s_{i-m_1+1,j-m_2+1} & \cdots & s_{i-m_1+1,j} \\ & \cdots & \\ s_{i,j-m_2+1} & \cdots & s_{i,j} \end{bmatrix}$$

where $m(= log_2(n+1)) = m_1 \cdot m_2$.

Example 7 $RP_{5,3} = (P_{5,3}, Q_p, D, I, R)$
where

$$P_{5,3} = \{p_{i,j} \mid 0 \le i \le 4, 0 \le j \le 2\}$$
$$Q_p = GF(2)$$

$$D = 1 + x \cdot y + x^2 \cdot y^2 + x^3 + y^2 +$$
$$x^2 \cdot y + x^2 \cdot y^2 + x \cdot y^2$$
$$I = \{0\}$$

and R is a mapping for a tree processor.

Labels for all processing elements in state "1" are

$$id_{0,0} = \begin{bmatrix} 0 & 1 \\ 0 & 1 \end{bmatrix} \quad id_{1,1} = \begin{bmatrix} 1 & 0 \\ 0 & 1 \end{bmatrix}$$

$$id_{0,2} = \begin{bmatrix} 0 & 0 \\ 0 & 1 \end{bmatrix} \quad id_{2,2} = \begin{bmatrix} 1 & 1 \\ 1 & 1 \end{bmatrix}$$

$$id_{3,0} = \begin{bmatrix} 1 & 1 \\ 0 & 1 \end{bmatrix} \quad id_{2,1} = \begin{bmatrix} 0 & 0 \\ 1 & 1 \end{bmatrix}$$

$$id_{1,2} = \begin{bmatrix} 0 & 1 \\ 1 & 0 \end{bmatrix} \quad id_{3,2} = \begin{bmatrix} 1 & 0 \\ 1 & 1 \end{bmatrix}$$

Using these labels, data can be transmitted from a source processing element to destination ones under the routing R. Fig.2 shows the tree processor implemented in $RP_{5,3}$.

The processor, RP_{s_1,s_2}, can be reconfigured by changing its state for setting the states of all faulty processing elements to the states in the set I as defined in Definition5. The processor based on a TDLCA satisfying the condition in Theorem1 can be reconfigured by making a state transition since the set of states for the processor is a subset of a minimal ideal of a polynomial ring generated by the two variable polynomial $W_2(x, y)$ and it has the CWED properties[10]. When an ideal is a minimal, the product of the weighting fuction and any polynomials in the ring is also an element of this subset. This means that the processor can be reconfigured by making a state transition from a certain state to the state in which the states for all of the faulty processing elements are additive unit element in GF(q). The processor state from which a transition is made and processor state after the transition in which state of all faulty processing elements are set to the additive unit is called the set of reconfiguration states. Thus, the set is defined as

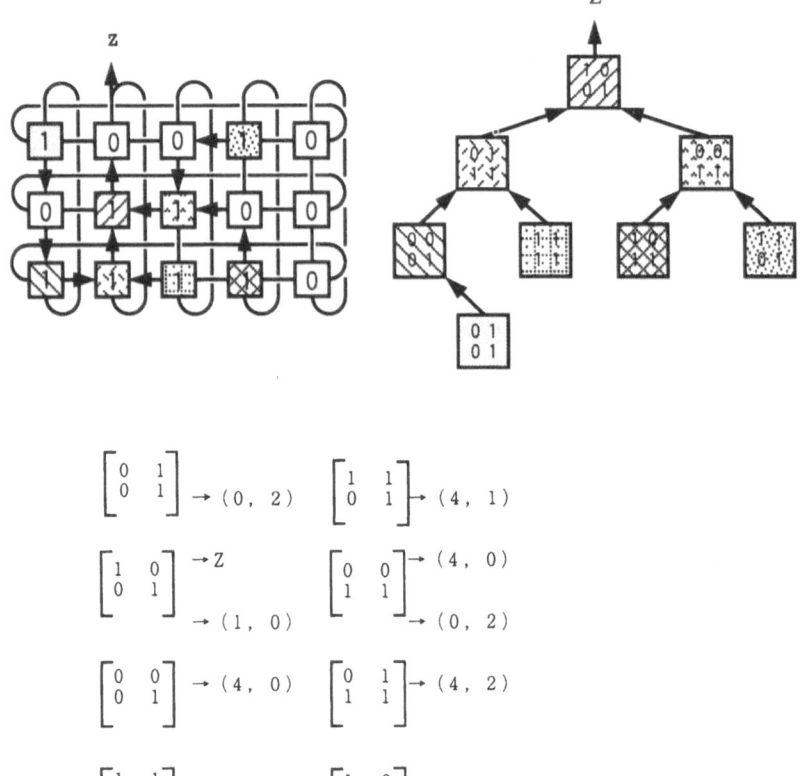

$$\begin{bmatrix} 0 & 1 \\ 0 & 1 \end{bmatrix} \rightarrow (0, 2) \qquad \begin{bmatrix} 1 & 1 \\ 0 & 1 \end{bmatrix} \rightarrow (4, 1)$$

$$\begin{bmatrix} 1 & 0 \\ 0 & 1 \end{bmatrix} \begin{matrix} \rightarrow Z \\ \\ \rightarrow (1, 0) \end{matrix} \qquad \begin{bmatrix} 0 & 0 \\ 1 & 1 \end{bmatrix} \begin{matrix} \rightarrow (4, 0) \\ \\ \rightarrow (0, 2) \end{matrix}$$

$$\begin{bmatrix} 0 & 0 \\ 0 & 1 \end{bmatrix} \rightarrow (4, 0) \qquad \begin{bmatrix} 0 & 1 \\ 1 & 1 \end{bmatrix} \rightarrow (4, 2)$$

$$\begin{bmatrix} 1 & 1 \\ 1 & 1 \end{bmatrix} \qquad\qquad \begin{bmatrix} 1 & 0 \\ 1 & 1 \end{bmatrix}$$

Figure 2: Tree Processor.

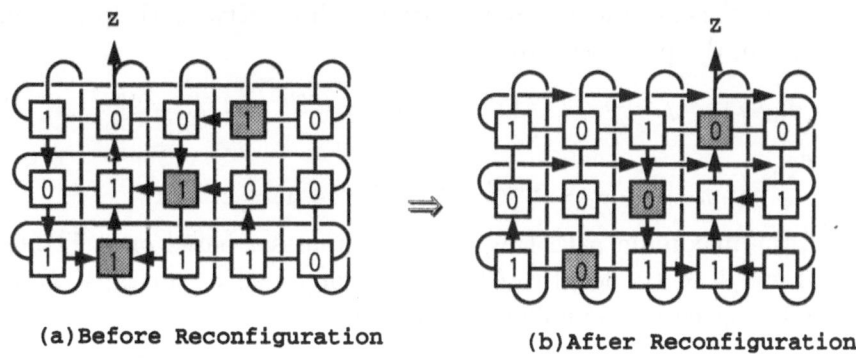

(a)Before Reconfiguration   (b)After Reconfiguration

Figure 3: Reconfiguration for a parallel processor.

Definition 6 *The set of reconfiguration states $F_{a,b}$ for a processing element $p_{a,b}$ is*

$$F_{a,b} = \{x^{i_1} \cdot y^{i_2} \mid w_2(x,y) \cdot x^{i_1} \cdot y^{i_2} \tag{8}$$
$$= s_{0,0} + s_{0,1} \cdot y + \cdots + s_{s_1-1,s_2-1} \cdot x^{s_1-1} \cdot y^{s_2-1}$$
$$0 \le i_1 \le s_1 - 1, 0 \le i_2 \le s_2 - 1, s_{a,b} = 0\}.$$

The number of elements in the set $F_{a,b}$ is $(n+1)/2-1$ since the number of additive unit elements in a state is always $(n+1)/2-1$ when q of GF(q) is 2. Moreover, the intersection between any r sets , $F_{a_1,b_1}, F_{a_2,b_2}, \cdots$ and F_{a_r,b_r} is not an empty set when $r \le log_2(n+1) - 1$[12]. From the above characteristics, the following lemma holds.

Lemma 1 *When faults are detected in processing elements, $p_{c_1,d_1}, \cdots ,$ and p_{c_r,d_r} ($r \le log_2(n+1) - 1$), the processor can be reconfigured by setting the state of the processor to a state in $F_{c_1,d_1} \cap \cdots \cap F_{c_r,d_r}$ and making a state transition.*

Fig.3 shows the reconfiguration of the processor shown in Example2.

3.4   Evaluation

The maximum path length L_1 between processing elements in a reconfigurable parallel processor based on one-dimensional linear cellualr automaton(LCA) model is represented by the following equation[12].

$$L_1 = n/2 = (2^m - 1)/2 \tag{9}$$

where n is the number of processing elements and m is a positive integer. On the other hand, the maximum path length L_2 between the processing elements of the proposed processor is

$$L_2 = s_1/2 + s_2/2 = (2^{m/2} - 1)/2 + (2^{m/2} + 1)/2 = 2^{m/2} \qquad (10)$$

where

$$n = s_1 \cdot s_2 = (2^{m/2} - 1) \cdot (2^m/2 + 1) = 2^m - 1.$$

Thus, the maximum path length of the processor based on a TDLCA model is $1/k$ of the length of the processor based on an LCA model, where k is given by the following equation.

$$1/k = L_1/L_2 = (2^{m/2} - (2^{-m/2})/2 \simeq 2^{m/2} = \sqrt{n} \qquad (11)$$

From the equation, we can say that the average path length of the processor based on a TDLCA model decreases in proportion to the square root of the number of processing elements compared with that of the processor based on an LCA model.

The processor proposed in this paper can operate normally when the number of faulty elements is less than $log_2(n + 1) - 1$, where n is the number of processing elements. The reliability of the processor proposed in this paper is compared with that of the non-redundant parallel processor and a duplex parallel processor, which uses about the same amount of hardware as the proposed processor. The reliability, $R_1(t)$, of the non-redundant processor is represented by Eq.(13), using the reliability of the local control part, $r_c(t)$, the reliability of the data processing part, $r_d(t)$, and $N(= \lfloor n/2 \rfloor)$, the number of processing elements in the processor. These reliabilities are assumed to be

$$r_c(t) = exp(-\lambda_c \cdot t),$$
$$r_d(t) = exp(-\lambda_d \cdot t),$$

where λ_c and λ_d are failure rate of a local control part and a data processing part, respectively.

$$R_1(t) = r_c(t)^N \cdot r_d(t)^N \qquad (12)$$

The reliability, $R_2(t)$, of the duplex parallel processor is represented by the following equation since the number of processing elements in the processor is $2 \cdot N$.

$$R_2(t) = 2 \cdot r_c(t)^N \cdot r_d(t)^N - r_c(t)^{2N} \cdot r_d(t)^{2N} \qquad (13)$$

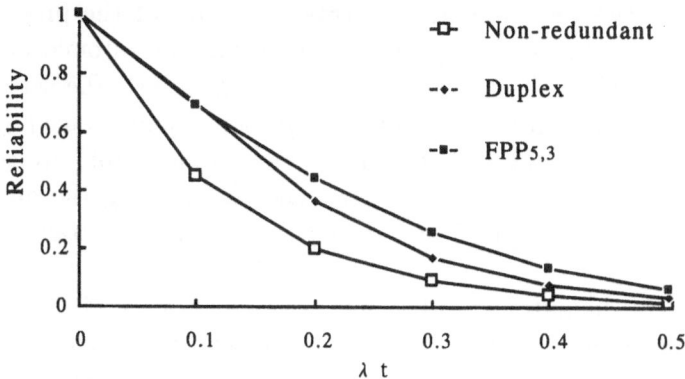

Figure 4: Reliability for a parallel processor.

The proposed parallel processor has $2 \cdot N - 1$ processing elements and it can operate normally even when $log_2(2 \cdot N) - 1$ data processing parts are faulty. When faults arise in the local control parts, every processing element has four data path to neighboring processing elements so that an active processing element can receive and transmit data when three local control part of the neighboring processing elements are faulty. Thus, the worst case reliability of the processor is represented by the following equation.

$$R_3(t) = r_c(t)^N \cdot \sum_{i=0}^{log_2(2 \cdot N) - 1} \binom{N-1}{i}$$
$$\cdot \ r_c(t)^{N-1-i} \cdot (1 - r_c(t)^i)$$
$$\cdot \ \binom{2N-i}{j} \cdot r_d(t)^{2 \cdot N - 1 - j} \cdot (1 - r_d(t)^j)$$

These three reliabilities are shown in Fig.4.

4   Conclusion

A reconfigurable parallel processor based on a two dimensional linear cellular automaton(TDLCA) model is proposed and a reconfiguration scheme for the processor utilizing the characteristics of the TDLCA is shown. In the model, the processing elements of the processor are regarded as the cells, and the operational states of the processor are treated as the states of the automaton. When an automaton used as the model satisfies the conditions given in Theorem1, the set of states of the automaton can have constant-weight and equidistance properties. This is the basic concept of the proposed reconfigurable processor. A reconfigurable parallel processor constructed by using the automaton model can be reconfigured quickly by making a state transition only once.

The states of the automaton having the properties are M-planes, therefore, allprocessing elements in the processor can have unique labels determined by the states of the neighboring processing elements. The connection between processing elements can be performed by routing R which determines the mapping from the source processing element to the destination ones. This connection can be changed easily by giving a different routing $R\prime$ to the processor for executing other algorithms.

The processor can be reconfigured by making the state transition only once even when multiple faults arise. Thus, the proposed parallel processor is quite useful for real-time data processing systems which require the high availability.

References

[1] R.W.Hockney and C.R.Jesshope, "Parallel Computers", *Adam Hilger Ltd.*, 1981.

[2] R.Duncan, "A Survey of Prallel Computer Architectures", *IEEE Computer*, vol.23, No.2, pp.5-16, 1990.

[3] M.Chea and J.Fortes, "A Taxonomy of Reconfigurable Techniques for Fault-Tolerant Processor Arrays", *IEEE Computer*, vol.23, No.1, pp.55-69, 1990.

[4] R.Negrini, M.G.Sami and R.Stefanelli, "Fault-Tolerance in VLSI and WSI Arrays", *MIT Press*, 1989.

[5] M.Tsunoyama and S.Naito, "A Fault-Tolerant Parallel Processor Modeled by A Linear Cellualr Automaton", *FTCS-88*, pp.334-339, June, 1988.

[6] M.Tsunoyama and S.Naito, "A Fault-Tolerant FFT Processor", *FTCS-91,* pp.128-135, June, 1991.

[7] R.E.Blahut, "Fast Algorithms for Digital Signal Processing", Reading M.A.:Addison-Wesley, 1984.

[8] M.Kawanaka, M.Tsunoyama and S.Naito, "A Fault-Tolerant Parallel Processor Modeled by A Two-dimensional Linear Cellualr Automaton", *25-th FTC-Workshop* 1992.

[9] van der Waerden, "Modern Algebra", *Springer Verlag,* 1940.

[10] M.Kawanaka, "A Fault-Tolerant Parallel Processor Based on a Two-dimensional Linear Cellualr Automaton", Master Thesis, Tech. Univ. of Nagaoka, 1992.

[11] M.Tsunoyama and S.Naito, "A Construction Method and Characteristics of Cellualr Automaton with Constant-weight and Equidistance Properties", *IEICE Trans.* vol.J70-D, No.12, pp.2348-2354, 1987.

[12] M.Tsunoyama and S.Naito, "A Fault-Tolerant Parallel Processor Based on a Linear Cellular Automaton Model", *IEICE Trans.* vol.J72-D-1, No.6, pp.491-497, 1989.

[13] J.Wakerly, "Error Detecting Codes, Self-Checking Circuits and Applications", *North-Holland,* NewYork, 1978.

[14] W.W.Peterson and E.J.Weldon, "Error-Correcting Codes", *MIT Press,* 1972.

[15] I.F.Blake and C.M.Ronald, "The Mathematical Theory of Coding", *Academic Press,* NewYork, 1975.

[16] L.N.Herstein, "Topics in Algebra", Wiley, NewYork, 1975.

A Modeling Approach for Dynamically Reconfigurable Systems

H. D. MEER* and H. MAUSER

Institute of Mathematical Machines and Data Processing IV

University of Erlangen-Nuremberg

Martensstr. 1, D-8520 Erlangen, Germany

Abstract

Based on Markov reward models we introduce a modeling approach which allows us to describe the behavior of dynamical systems. We present an algorithm which evaluates and optimizes rather general configurations. We choose the expected cumulative reward over a finite time horizon as the optimality criterion. The method is applied to control and evaluate responsive queueing systems. The models are evaluated in the presence of subcomponent failures.

Key Words: Performability analysis, adaptive systems, responsive queueing systems, Markov decision processes, extended Markov reward models.

1   Introduction

There has recently been growing interest in such concepts as dynamic adaption of real-time software [1], adaptive fault tolerance [2], or unifying fault and change management for distributed systems [3]. In real-time systems, such as flight control systems, a timely response has to be provided together with a high reliability level [4]. Often the environmental demands for service change over time due to peaks in the computational load or evolutionary changes in external conditions. Furthermore system components may fail or be added. Thus it is tempting

---

*Now with Department of Electrical Eng., Duke University, Durham, N.C. 27708, U.S.A.

to develop methods for efficiently and *adaptively* utilizing a limited and *dynamically* changing amount of resources. Though research is being pursued towards defining and establishing the notion of adaptive and dynamically changing systems, modeling and evaluation of such systems and algorithms has hardly been done. Note that typically transient performance measures provide more meaningful information for evaluation and control of the systems we are interested in than steady state measures do.

Reconfiguration is an important concept in modeling and evaluation of fault tolerant systems. For instance, a degradable system may be reconfigured to a less effective but still operational mode as a result of a failure which is, in general, a random event. We pursue a more general notion of reconfiguration. According to our approach it takes place not only in response to faults, but can occur at any instant of time at which the system is in one of a *prespecified* set of states. Thus reconfiguration due to a fault is included as a special event. Reconfiguration in the broader sense is performed by an initiated and controlled switch between distinguished states in order to optimize a certain performability measure [5], i.e., *the expected accumulated reward, $E[Y_z(t)]$, in the mission interval t*. The intention is to enhance the overall system performance. Examples of events that could trigger a reconfiguration include a change of the current load, failures of system components, or, less obviously, the mere passage of time. The latter issue is particularly important in real time applications. In general several possible reconfigurations have to be evaluated in a certain state and the optimal decision may change as a function of time. The most appropriate compromise among different levels of performance, each of which can be achieved with its attendant expense, should be determined.

In connection with performability evaluation there is a significant lack of optimization studies. Expected time averaged reward coupled with repair cost penalties is used in [6] as a criterion of optimization. The method is based on semi-Markovian decision processes and is applied to steady state and periodic models. They apply the well-known policy improvement procedure for the optimization. Related work has been done using techniques of sensitivity analysis [7, 8, 9]. However, these methods are restricted to static optimization. Our approach is based on semi-Markov decision processes with the expected accumulated reward as the optimization criterion [10]. We investigate dynamic optimization and evaluation of reconfigurable systems via *Extended Markov Reward Models* (EMRM). Since the concept of EMRM allows us to integrate the computation of a performability measure, i.e., $E[Y_z(t)]$, and the optimization of the evaluated system, we regard it as a new concept. EMRM augment Markov reward models, i.e., continuous time Markov chains (CTMC), where reward rates are associable with states, with the option of *optimization* coupled with *state transitions* to which *rewards* may be associated.

As the method of analysis we introduce a new algorithm, which is based on a numerical solution of the system of integral equations which describe the transient behavior of the system being investigated. All features mentioned above may essentially be captured by the same unifying technique for computing $E[Y_z(t)]$. The algorithm proceeds backwards in time, a similar technique as proposed independently in [8], where partial differential equation methods and piecewise deterministic Markov processes are applied. No dynamic optimization is performed in [8]. We augment former work of Lee and Shin [11], who deal with optimizing a non-repairable configuration, i.e., an acyclic model structure, to a general framework. Acyclic **and** non-acyclic Markov models with a rather sophisticated reward structure attached may be evaluated. Although not addressed in the current paper, non-homogeneous Markov reward models can also be analyzed by our method.

Subsequently we introduce the concept of Extended Markov Reward Models (EMRM). We proceed by presenting the details of the computational algorithm. In section 3 the method is applied to control and evaluate responsive queueing systems in the presence of subcomponent failures.

2   Extended Markov Reward Models

2.1   The Basic Features

Markov reward models (MRM) are a powerful tool for modeling dependability and performance of fault tolerant systems. Particularly for the combined evaluation of both characteristics MRM are extensively used [9, 12, 13]. In the following we are concerned both with the evaluation and the optimization of Markov models. In our approach the computation steps are performed backwards in time, i.e., a time-to-go approach.

As one extension of the general structure of the MRM, *rewarded transitions* are admissible in our approach. Thus *pulse rewards* are a useful modeling tool. The most important feature of EMRM is the possibility of specifying optional *reconfigurations* as an integral part of the Markov model. An example of an EMRM is shown in Figure 1. Markov states

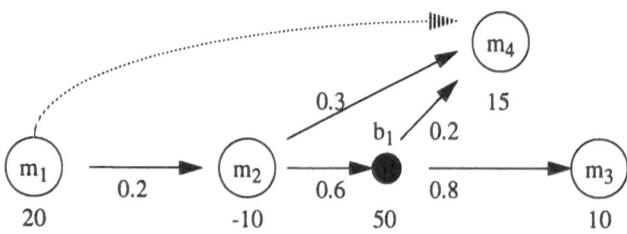

Figure 1: Example of an EMRM

m_i are represented by circles and (rewarded) state transitions b_i are represented by dots. For convenience we call them *branching states*. The arcs leaving Markov states are labelled with rates, and those leaving branching states with probabilities. The numbers attached to the states denote the corresponding reward rates. The possibility of a reconfiguration is indicated in the figure by a dotted arc. The interpretation is as follows: Whenever the system is in state m_1 a decision is made whether to reconfigure the system to state m_4 or not. The most promising alternative, i.e., the one which yields the highest expected reward in a given mission time is chosen. Obviously one would reconfigure to state m_4 in the long run. For shorter mission times the situation is more complicated. Does the chance to receive the pulse reward associated with state b_1 trade-off the cost which has to be paid for passing through state m_2? Furthermore, under which conditions does the pulse reward compensate for the risk of reaching m_3 eventually? [1]

The dotted arc in Figure 1, or equivalently the corresponding reconfiguration, represents a feature of EMRM which is regarded as an *application*

---

[1] m_4 outperforms m_3 because of the higher reward rate which is attached to it.

oriented abstraction of the underlying Markov decision theory. The user of EMRM does **not** have to be familiar with the complex details of Markov decision theory; rather he can think in terms of the application context and directly specify the model in terms of the interesting reconfiguration problem. This feature is provided by a marriage between performability modeling and dynamic programming techniques.

2.2 The Mathematical Concept

We assume (negative) exponentially distributed holding times of the *Markovian* states in our models. For example, the CDF of a random variable T_0, which characterizes the holding time of a state z_0, is computed as

$$F_{T_0}(t) = 1 - e^{-\sum_{i=1}^{n} q_{0,i} t}.$$

Thus the holding time is exponentially distributed with parameter $\sum_{i=1}^{n} q_{0,i} = q_0$, which is the total rate of leaving z_0. The probability of transitions from state z_0 to z_i is defined, for all i, by $p_{0,i} = \frac{q_{0,i}}{q_0}$, where q_0 is defined as above, and $q_{0,i}$ represents the state transition rate from state z_0 to z_i.

Associating real valued reward rates with the Markov states leads to the concept of *Markov reward models*. The behavior of a system over time is characterized by the stochastic process $\{Z : \Re^+ \longrightarrow S\}$, where S comprises the finite set of all possible states. $\{Z(t); t \geq 0\}$ is often called the *structure state process*. The random variable $Z(t)$ denotes the state of the underlying process at time t. To indicate the *instantaneous reward rate* at time t and the *accumulated reward* in the interval $(0,t)$, we use the random variables $X(t)$ and $Y(t)$. $Z(t), X(t)$, and $Y(t)$ are non-independent random variables. $X(t) = r(Z(t))$, where r denotes the reward rate associated with the current state $Z(t)$, and $Y(t) = \int_0^t X(\tau)d\tau$. The distribution of $Y(t)$, i.e., $P(Y(t) \leq y)$ for any reward y, is called the *performability* [5].

We are interested in computing $E[Y_z(t)]$, the expected accumulated reward conditioned on the initial state z, for EMRM. Later in this paper we will use this measure as a criterion of optimization. $E[Y_z(t)]$ can be computed for all states z by the following system of integral equations. Note that we have to distinguish between Markov and branching states. For both cases we define $E[Y_z(0)] \equiv 0$.

Markov States:

$$E[Y_z(t)] = r(z)te^{-q_z t} + \int_0^t q_z e^{-q_z \sigma}(r(z)\sigma + \sum_{i=1}^{n} p_{z,z_i} E[Y_{z_i}(t - \sigma)])d\sigma \quad (1)$$

$e^{-q_z t}$ represents the probability that the system, started in state z, will
not make a transition out of that state during $[0, t]$. Thus, the total
reward gained is $r(z) \times t$. This is reflected by the first term of the equation. The integral indicates the other event, namely that a state change
occurs during the interval $[\sigma, \sigma + d\sigma], 0 \leq \sigma, \sigma + d\sigma < t$. In this case
the mean accumulated reward is composed of two weighted parts: the
reward gained in the state z, $r(z) \times \sigma$, plus the mean reward achieved
in the remaining mission time, $t - \sigma$, depending on the successor states
of the transition.

Branching States:

$$E[Y_z(t)] = r(z) + \sum_{i=1}^{n} p_{z,z_i} E[Y_{z_i}(t)] \quad (2)$$

Since no time is spent in a branching state, the remaining mission time
is t. But the constant reward $r(z)$ is accumulated in any case. We now
present a numerical solution method of the above system of recursive
integral equations.

2.3   The Method of Computation

The solution method is based on discretization of the finite time horizon. The interesting measures are computed backwards in time. The
accuracy of the algorithm depends on the size of the time steps Δt. One
has to choose Δt such that the probability that more than one state
transition occurs in Δt approaches zero. As a default value Δt is chosen
as approximately a hundredth of the mean state holding time. Since Δt
can be chosen adaptively, the computational overhead can be reduced
to a minimum at each computation step.
For all time steps and for all states the following basic computations have
to be performed. We define the vector $\underline{E}[Y_z(t)] = (E[Y_z(t)], E^1[Y_z(t)]$
$, \ldots, E^m[Y_z(t)])$, where $E^m[Y_z(t)]$ denotes the m-th derivative of $E[Y_z(t)]$.
For each time step $\underline{E}[Y_z(t + \Delta t)]$ can be computed iteratively by using

the quotient of differences as an approximation:
$E^m[Y_z(t + \Delta t)] \approx \frac{E^{m-1}[Y_z(t+\Delta t)]-E^{m-1}[Y_z(t)]}{\Delta t}$, where $m \geq 1$ and $E^0[Y_z(t)] = E[Y_z(t)]$. Assume $E[Y_z(t)]$ has already been computed for all $z \in S$ and a certain $t < T$, where T denotes the finite time horizon. Now using *Taylor polynomials* of order m as a method of approximation the further evaluation of $E[Y_z(t)]$ in a small time interval Δt can be interpolated. Thus, the above recursive integrals transform into definite integrals, which may be computed easily. Concerning the current time step, Eq. 1 can be rewritten as Eq. 3 below. For illustration of a time step see Figure 2.

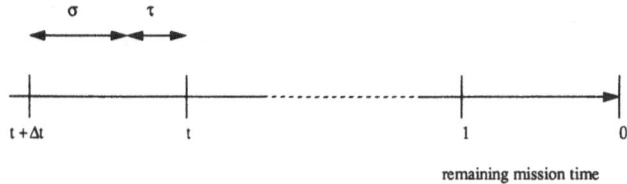

Figure 2: System evolution during a small increment of time

Markov States:

$$E[Y_z(t + \Delta t)] = e^{-q_z \Delta t}(r(z)\Delta t + E[Y_z(t)])$$
$$+ \int_0^{\Delta t} q_z e^{-q_z \sigma}(r(z)\sigma + \sum_{i=1}^{n} p_{z,z_i} E_{pol}[Y_{z_i}(t + \Delta t - \sigma)])d\sigma \qquad (3)$$

We use the Taylor polynomial to define: $E_{pol}[Y_z(t + \tau)] = E[Y_z(t)] + E^1[Y_z(t)]\tau + \cdots + E^m[Y_z(t)]\frac{\tau^m}{m!}$
Having computed $E[Y_z(t + \Delta t)]$ for all Markov states, the branching states remain to be evaluated.

Branching States:

$$E[Y_z(t + \Delta t)] = (r(z), \{0\}^m) + \sum_{i=1}^{n} p_{z,z_i} E[Y_{z_i}(t + \Delta t)] \qquad (4)$$

Again m denotes the order of the Taylor polynomial, and $\{0\}^m$ represents a m-tuple with 0 components. Note that $E[Y_{z_i}(t+\Delta t)]$ of possible successor states z_i of z have to be computed first. The recursion terminates when the first Markov state in that row is evaluated.

Eq. 3 and 4 are the main equations. For the ease of implementation we include the following simplications. With some algebra Eq. 3 can be evaluated and transformed into the simple formula shown in Eq. 5.

$$E[Y_z(t + \Delta t)] = e^{-q_z \Delta t}(r(z)\Delta t + E[Y_z(t)]) + \sum_{i=0}^{m} k_i I_i \qquad (5)$$

For $m = 3$ the coefficients k_i evaluate to:

$$k_0 = \overline{E} + \overline{E^1}\Delta t + \frac{1}{2}\overline{E^2}\Delta t^2 + \frac{1}{6}\overline{E^3}\Delta t^3$$

$$k_1 = r(z) - \overline{E^1} - \overline{E^2}\Delta t - \frac{1}{2}\overline{E^3}\Delta t^2$$

$$k_2 = \frac{1}{2}\overline{E^2} - \frac{1}{6}\overline{E^3}\Delta t$$

$$k_3 = -\frac{1}{6}\overline{E^3}$$

In the above equations the following abbreviation is used: $\overline{E^m} = \sum_{i=1}^{n} p_{z,z_i} E^m[Y_{z_i}(t)]$. For I_i one finds:

$$I_0 = 1 - e^{-q_z \Delta t}$$

$$I_1 = \frac{I_0}{q_z} - \Delta t e^{-q_z \Delta t}$$

$$I_2 = 2\frac{I_1}{q_z} - \Delta t^2 e^{-q_z \Delta t}$$

$$I_3 = 3\frac{I_2}{q_z} - \Delta t^3 e^{-q_z \Delta t}$$

Observe that normally Taylor polynomials of first order, $m = 1$, yield satisfying results for our application. This results in an efficient algorithm. The largest models we investigated with a reasonable computation time were in the order of hundreds of thousands of states. A sparse storage technique and an adaptive selection of Δt makes the computation even more efficient. Hence, also stiff models can be evaluated.

Recall Eq. 2 for the computation of $E[Y_z(t+\Delta t)]$ of the branching states. Now let us assume a given Markov state z_0 and, additionally to a possibly underlying Markov structure, potential successor states $z_1, z_2, \ldots, z_n$, which are optimization options. In other words $z_1, z_2, \ldots, z_n$ are all states to which the system can be reconfigured whenever it is in state z_0. Then the optimization is performed by determining the solution θ of Eq. 6 at each time step Δt.

$$E[Y_\theta(t + \Delta t)] = \max\{E[Y_{z_i}(t + \Delta t)] | z_i \in \{z_0, \ldots, z_n\}\} \tag{6}$$

In [14] it is shown how the solution of Eq. 6 corresponds to the theory of Markov decision processes. We avoid the lengthy discussion of this topic here. To keep things simple the optimization is illustrated by using the basic features of EMRM only. Thus the effect of an optimization decision can be thought of as a transformation of z_0 into a branching state with deterministic transition to z_i, i.e., $p_{z_0, z_i} = 1$, if z_i is found to maximize Eq. 6 and $z_i \in \{z_1, \ldots, z_n\}$. If z_0 maximizes Eq. 6 the possibilities of reconfiguration are simply neglected. Note that by choosing larger time steps the interpolation scheme above allows also a fast approximative computation of the reward and strategy with an high accuracy.

3 An Example: Control in the Presence of Failures

3.1 The Basic Model

In real-time systems a certain guarantee to deliver the required service in time is mandatory. The guarantee may be provided on a per job basis or, more realistically, as a probability measure that the response time is within an acceptable range. A common practice in control theory is to determine a threshold on the number of jobs that can be admitted to the system under investigation. The threshold defines the maximum number of jobs that may be in the system at the same time. With a sufficient high probability the response time can be assured for these customers. Any additionally arriving customer must be rejected. As an appropriate modeling approach *finite capacity queueing systems* are extensively used for this kind of problem [15, 16]. Note that we are not concerned with constructing and proving optimal thresholds. In the following we assume them as given.

The scenario we are interested in is more complicated in the sense that we wish to consider the case when the resources are subject to failures. This can be regarded as another level of modeling and control superimposed over the above mentioned problem. The modeling and evaluation

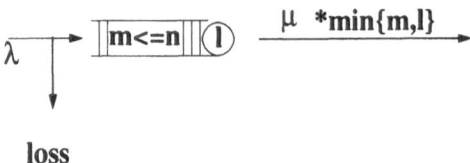

loss

Figure 3: A finite capacity queueing system subject to server breakdown

of real-time system performance in the presence of failures has been
studied in [7]. In this related work also performability measures are
used to evaluate the combined effects of performance, dependability,
and deadline violations. But control is not considered. For real time
systems it is tempting to provide a unified methodology which allows
the evaluation and control of the system. It is extremely important to
react immediately if, for example, a component failure occurs. An action should be taken which minimizes the number of jobs which are lost
or, equivalently, which are not delivered in time. The failed component
may be repaired or the load may be redirected to other resources for
service. Obviously, the usefulness of a specific measure depends on the
current state of the system and on the time.

Figure 3 depicts the basic structure of a dependable queueing system.
The system is assumed to be composed of a queue of finite capacity
$n = 10$. l servers, $l \leq i = 5$, are functioning properly. We assume that
only the servers are unreliable; the buffer elements are failure free. A *loss*
occurs if the buffer is full, i.e., $m = n$ where m denotes the number of
customers in the system. The underlying assumption is that n is chosen
such that the response time is guaranteed for any value of l.[2] For each
lost customer a pulse reward -1 results. A reward rate 0 is attached
to all other states. Maximizing $E[Y_z(t)]$ results in the minimization of
the mean number of losses. For simplicity we assume a 'first come first
served' service strategy. Each server has the service rate $\mu = 0.5$. The
jobs arrive to the system with rate $\lambda = 1$.

The service capacity of the system is governed by a failure/repair pro-

---

[2]Alternatively, a straight forward extension to the model would be to require a minimum service capacity $\bar{l} > 0$ for the system being up. If $l < \bar{l}$ all present customers would
be discarded then too.

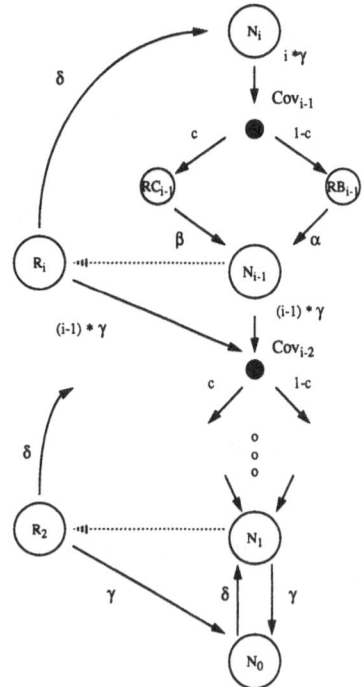

parameter	meaning
i	max. capacity
c	coverage prob.

transition rate	meaning
γ	failure rate
δ	repair rate
α	reboot rate
β	recovery rate

state	meaning	type
N_j	j failure free elements	Markov(D)
RC_k	recovery to N_k	Markov
RB_k	reboot to N_k	Markov
Cov_k	coverage to N_k	branching
R_l	repair from N_{l-1} to N_l	Markov

Figure 4: A reconfigurable multi level system with time redundancy

cess. A model of the structure state process is shown in Figure 4. In Markov state N_l, $0 \leq l \leq i$, l sever units are functioning properly. Each server may fail with rate γ. If a component failure occurs the system may recover successfully with coverage probability $c = 0.9$ to a less effective but still operational level N_{l-1} if at least one server remains functioning. With probability $1 - c = 0.1$ a reboot is necessary. Reboot and recovery rates are chosen as $\alpha = 0.5$ and $\beta = 5$, respectively. The system is considered to be *down*, and hence an arriving customer is *lost*, if no server is functioning, i.e., $l = 0$, or the system is being rebooted or reconfigured. As mentioned above, an arriving customer is also lost if the buffer is *full*, i.e., $m = n$.

Recovery operations, reboot operations and performance degradation (reconfiguration) are kinds of time redundancy. On the one hand, the higher the service capacity is the less likely a loss occurs due to buffer overflow. On the other hand, the total failure rate $l * \gamma$ increases linearly with the number l of operational components; the higher the total failure rate is the more likely a loss occurs due to recovery or reboot.[3] Obviously there exists an optimal level of redundancy, i.e., service capacity. The optimum is load and time dependent.

To keep things simple we pursue merely the control of the repair process. Of course more involved reconfiguration schemes can be investigated. We are primarily interested in providing adequate responses to events which cause a change of the load in the system or a change of the structure of the system. We assume one repairman being available who can repair a single component with rate δ. The repairman has the opportunity to refuse the repair action if the system would benefit from this decision. The criterion of optimization is the expected number of losses. Since the arrival rate λ is set to 1 this measure equals the mean *interval unavailability* of the system to customers. All stochastic assumptions are of Markovian nature.

The dotted arcs in Figure 4 represent the possible repair decisions. Whenever the system is in *structure state* N_l it can be reconfigured to the corresponding repair structure state R_{l+1} for on-line repair. Note that each structure state N_l, say, corresponds to a class of system states $N_{m,l}$, where m, $0 \leq m \leq n$, denotes the number of customers in the system. The overall model is composed of the performance model in figure 3 and the structure state model in Figure 4.

3.2   The Results

In this section we discuss selected state dependent control strategies and reward functions. As parameters of interest the failure rate γ and the repair rate δ are chosen.

For each fixed set of parameters and each state $N_{m,l}$ in which an optional reconfiguration is possible a strategy encompasses an infinite set of time dependent optimal decisions. The strategies are represented in a compact way by so called *switching curves*. The curves mark the instants of time where decisions change. Each switching curve is exactly related to one Markov decision state and a fixed value of the failure rate γ.[4] The

---

[3]Note that the impact of a failure event can even be worse. For example, if a reboot takes too long all present customers have to be discarded.

[4]Markov decision states are characterized in the tables attached to Figure 4 by the type 'Markov(D)'.

numbers attached to the time axes of Figures 5, 6, 7, and 8 denote the *remaining mission time*. Hence time passes "from the top of the figures to the bottom". The abscissa marks the end of the mission time.

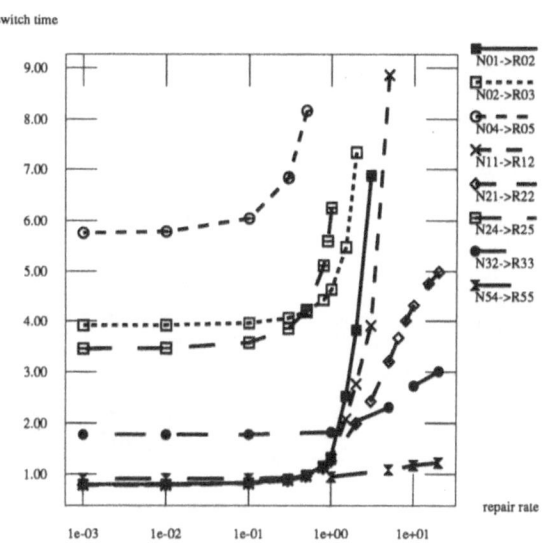

Figure 5: Strategy switch for a responsive queueing system

In Figure 5 the optimal strategies for selected system states are depicted. The results are computed for a time horizon of 100. State $N_{m,l}$ denotes the case where m customers are in the system while l servers are functioning. The failure rate γ is set to 10^{-4} and the repair rate δ is varied from 10^{-3} to 20. The curves indicate the switch time of the strategy with respect to the remaining time horizon as a function of the repair rate. For example, in state $N_{1,1}$ repair is initiated for small repair rates, e.g., rates less than 10^{-2}, until almost the end of the time horizon. In this case the strategy switches at 0.8 units of time from *'do repair'*, i.e., do reconfigure to the respective repair state $R_{1,2}$, to *'do not repair'*. Hence, with respect to the remaining mission time, repair is performed "above" the switching curve; "underneath" repair is not done. If the repair rate comes close to 1, the strategy for $N_{1,1}$ becomes very sensitive to a further increase. For a repair rate of 5, repair will only be performed if the remaining time horizon is more than 8.859 units of time.

A further increase of the repair rate to 6.5 leads to the decision not to
repair at all in the evaluated time horizon of 100.

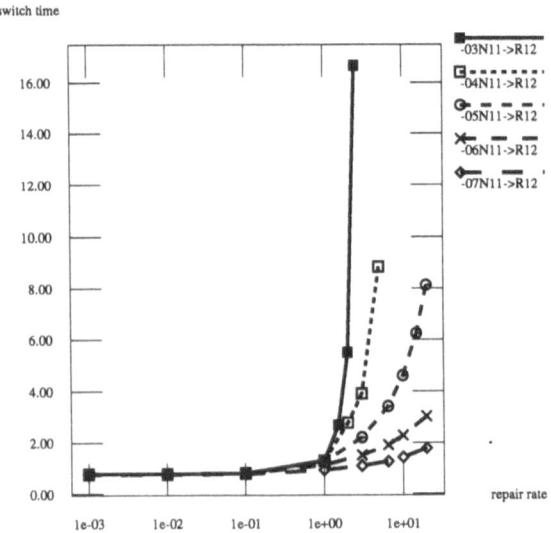

Figure 6: The influence of failure and repair rates in the case of one
customer and one server

Both the influence of the load and the available resources on the decisions
can be concluded from the figure. For instance, if one server is available
there are hardly any differences in the strategies if the repair rate is less
than 1. For higher repair rates the load has an increasing impact on
the strategy, as is seen by comparing the switching curves for $N_{0,1}, N_{1,1}$,
and $N_{2,1}$. Furthermore, the similarity of the curves of these states for
small rates is a special case. This comes from the fact that a failure
of the only resource leads to a down state. This is not true for the
other configurations. Comparing $N_{0,4}, N_{2,4}$, and $N_{5,4}$ indicates that the
strategies differ significantly for small repair rates as well if the load is
varied. While the switching curve for $N_{5,4}$ is hardly sensitive to the repair
rate and repair is performed in almost every case, it is quite sensitive
for $N_{0,4}$. In this case repair will only be performed in the evaluated
mission interval if the repair rate is less than 0.8. The corresponding
switching curve $N_{0,4} \rightarrow R_{0,5}$ approaches infinity between two consecutive

sample points . The sensitivity of the strategy to the load is typical for responsive systems.

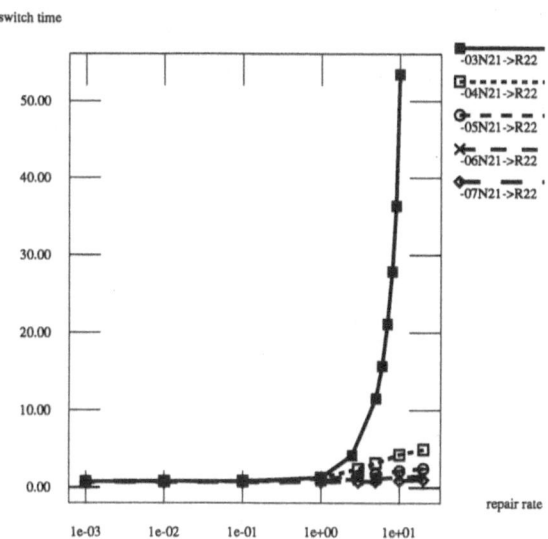

Figure 7: The influence of failure and repair rates in the case of two customers and one server

Figures 6, 7, and 8 indicate the dependence of the strategy upon the failure rate. The figures show the switching curves for the states $N_{1,1}, N_{2,1}$, and $N_{0,2}$ for different failure rates $\gamma = 10^{-7}, 10^{-6}, 10^{-5}, 10^{-4}$, and 10^{-3} as a function of the repair rate. Again, it is interesting to note the special case of the configuration where only one server is available. For repair rates less than 1, the switching curves are quite close together for all different failure rates. This is not true for the configurations where at least two servers are available. Here the strategy is sensitive to the failure rate for small repair rates as well. Furthermore, for repair rates greater than 1 the strategy is quite sensitive to the load also in the one server case. This can be seen by comparing the properties of the switching curves of states $N_{1,1}$ and $N_{2,1}$ in Figures 6 and 7.

It is to be noted that the higher the failure rates of the components are, the less repair will be initiated. This can be seen in all four strategy figures. It reflects the risk inherent in failure prone components.

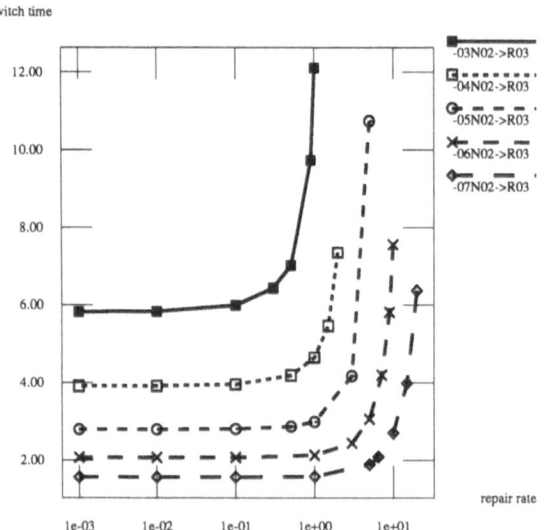

Figure 8: The influence of failure and repair rates in the case of zero customers and two servers

There is obviously a trade-off in the processing capacity and the system unavailability due to reconfiguration and reboot. Finding the optimal compromise is especially important for responsive systems.

In the underlying model of the *reward functions* which are depicted in Figure 9 the component failure rate is chosen as 10^{-3}. The reward functions show the mean number of losses in 100 units of time or, equivalently, the interval unavailability. Figure 9 relates to the corresponding switching curve of state $N_{1,1}$ and the same failure rate in Figure 6. ¿From Figure 6 it is evident that *no repair* should be performed if the repair rate is larger than 2. For repair rates $1 \leq \delta \leq 2$ the strategy is strongly dependent on time; repair is only performed in the long run. Figure 9 reveals the loss of performance if no optimizing decision (nd) is done. Without control repair would be performed indiscriminately.

Two conclusions can be drawn. Given the initial state $N_{1,1}$, which is chosen as an example, the figure indicates that a significant reduction in the total expected loss results if a decision process is incorporated. This example validates that a carefully controlled system can in effect

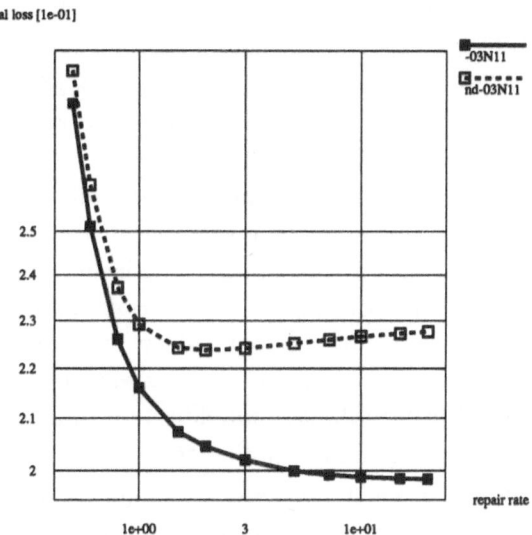

total loss [1e-01]

-03N11
nd-03N11

2.5
2.4
2.3
2.2
2.1
2

repair rate

1e+00   3   1e+01

Figure 9: To decide or not to decide?

enhance the system performance. Note that the only decision is to refuse repair in certain cases, which leads to a performance improvement up to $10 - 15$ percent! For the second implication one has to investigate the dotted curve carefully. Initially, as expected, the unavailability decreases as the repair rate increases. Surprisingly the reward (cost) function has its minimum at the repair rate 2, and then slightly increases again. One might be misled by concluding that the system would benefit from a 'lazy' repair in certain cases. But the true reason for this 'counter intuitive' behavior comes from neglecting the strategy and doing repair immediately in any case. In this case the risk of losing a customer due to recovery or reboot compensates for the possible gain in performance due to an additional server.

We are mainly interested in the strategies as a *function of time*. The properties of the computed policies may as well be used as an indicator of properties of an optimal value function. Some form of monotonicities can be used to prove convergence to *stationary* strategies in an infinite horizon [17]. Sample path techniques are used in [18] alternatively. But we do not consider this topic in depth here.

4   Conclusions

We presented a time-to-go approach to compute the mean cumulative reward $E[Y_z(t)]$ in a finite time horizon and used this measure as an optimization criterion. *Extended Markov Reward Models* (EMRM) were introduced as a unifying framework for evaluation and optimization of multi mode computer systems. EMRM are an outcome of a marriage between performability modeling and dynamic programming techniques. The most important feature of EMRM is the *optimization abstraction*. Users have not to be familiar with the underlying Markov decision theory. Rather they can specify an application oriented reconfiguration problem as an integral part of an ordinary Markov reward model. Markov reward models are used extensively for performance, dependability, and performability modeling and evaluation.

The computational method was applied to models of fault tolerant systems with degradable performance and finite queueing capacity. The results reflected the transient behavior of the configurations investigated. Our method allows simultaneous evaluation and control of reconfigurable systems. Thus insight into dynamic system behavior can be obtained and optimal dynamic control can be performed by an efficient table look up at run time. The efficiency of the method allows to investigate rather large models with hundreds of thousands of states.

Acknowledgements

We wish to thank H. Levy, J. Muppala, and K. Trivedi for many valuable discussions.

References

[1] T. A. Bihari, K. Schwan, "Dynamic Adaption of Real-Time Software", *ACM Trans. on Computer Systems*, Vol. 9, No. 2, (May 1991), pp. 143–174.

[2] K. H. Kim, T. F. Lawrence, "Adaptive Fault Tolerance: Issues and Approaches", *Proc. of 2nd IEEE Workshop on Future Trends of*

Distributed Systems, (Sept.30 - Oct.2 1990), Cairo, Egypt, PP. 38–46.

[3] J. Kramer, J. Magee, A. Young, "Towards Unifying Fault and Change Management", *Proc. of 2nd IEEE Workshop on Future Trends of Distributed Computing Systems*, (Sept.30 - Oct.2 1990), Cairo, Egypt, pp. 57–63.

[4] M. Malek, "Responsive Systems: A Marriage between Real Time and Fault Tolerance", *Proc. of 5th Intern. GI/ITG/GMA Conf. on Fault-Tolerant Computing Systems*,(Sept.25 - 27 1991), Nuremberg, Germany, pp. 1–17.

[5] J. F. Meyer, "On Evaluating Performability of Degradable Computing Systems", *IEEE Trans. on Computers*, Vol. C-29, No. 8, (1980), pp. 720–731.

[6] K. G. Shin, C. M. Krishna, Y. H. Lee, "Optimal Dynamic Control of Resources in a Distributed System", *IEEE Trans. on Software Engineering*, Vol. 15, No. 10, (1989), pp. 1188–1197.

[7] J. K. Muppala, *Performance and Dependability Modeling Using Stochastic Reward Nets*, Ph.D. thesis, Department of Electrical Engineering, Duke University, Durham, U.S.A., (1991).

[8] K. R. Pattipati, Y. Li, and H. A. P. Blom, "On the Instantaneous Availability and Performability Evaluation of Fault-Tolerant Computer Systems", *Proc. of IEEE Int. Conf. Syst., Man, Cybern.*, MA, (Nov. 1989), pp. 376–382.

[9] A. L. Reibman, R. S. Smith, and K. S. Trivedi, "Markov and Markov reward model transient analysis: An overview of numerical approaches", *European Journal of Operations Research*, Vol. 40 (1989), pp. 257–267.

[10] R. A. Howard, *Dynamic Probabilistic Systems, Vol. II: Semi-Markov and Decision Processes*, John Wiley & Sons, New York, (1971).

[11] Y. H. Lee, and K. G. Shin, "Optimal Reconfiguration Strategy for a Degradable Multimodule Computing System", *Journal of the ACM*, Vol. 34, No. 2, (1987), pp. 326–348.

[12] A. L. Reibman, K. S. Trivedi, "Transient Analysis of Cumulative Measures of Markov Model Behavior", *Commun. Statist. - Stochastic Models*, Vol. 5, No. 4, (1989), pp. 683-710.

[13] E. De Souza e Silva, H. R. Gail, "Calculating Availability and Performability Measures of Repairable Computer Systems Using Randomization", *Journal of the ACM*, Vol. 36, No.1, (1989), pp. 171–193.

[14] H. de Meer, *Transiente Leistungsbewertung und Optimierung rekonfigurierbarer fehlertoleranter Rechensysteme* (in German), Arbeitsberichte des IMMD (Informatik), Vol. 25, No. 10, University of Erlangen-Nuremberg, Erlangen, (1992).

[15] T. G. Robertazzi, A. A. Lazar, "On the Modeling and Optimal Flow Control of the Jacksonian Network", *Performance Evaluation*, Vol. 5, (1985), pp. 29–43.

[16] F. C. Schoute, "Overload Control in SPC Processors", *Philips Telecommunication Review*, Vol. 41, (1983), pp. 300–310.

[17] B. Hajek, "Optimal Control of Two Interacting Service Stations", *IEEE Trans. on Autom. Control*, Vol. AC-29, (1984), pp. 491–499.

[18] P. P. Bhattacharya, A. Ephremides, "Optimal Scheduling with Strict Deadlines", *IEEE Trans. on Autom. Control*, Vol. AC-34, (1989), pp. 721–728.

Author Index

Dependable Computing
and Fault-Tolerant Systems

Edited by A. Avizienis, H. Kopetz, J.-C. Laprie

J. F. Meyer, R. D. Schlichting (eds.)
Vol.6: Dependable Computing for Critical Applications 2

1992. 114 figs. XIV, 439 pages.
Cloth DM 172,–, öS 1204,–
ISBN 3-211-82330-1

The book contains the papers presented at the 2nd IFIP Working Conference on Dependable Computing for Critical Applications held in Tucson, Arizona, on February 18-20, 1991. Based on feedback at that meeting, these papers were then revised and updated prior to inclusion in this volume. The topics addressed span the spectrum of dependable computing, from design methods for distributed, fault-tolerant systems to formal and experimental validation techniques. The unique focus of this forum on critical applications is what distinguishes many of these papers from those found elsewhere. This book is of interest to individuals involved in the development of computing systems where dependability attributes such as reliability, safety, and security are a major concern.

J.-C. Laprie (ed.)
Vol.5: Dependability: Basic Concepts and Terminology
In English, French, German, Italian, and Japanese

1992. 3 figs. XII, 265 pages.
Cloth DM 128,–, öS 896,–
ISBN 3-211-82296-8

This volume gathers a text resulting from several years of work per-formed within the IFIP WG 10.4 on Dependable Computing and Fault Tolerance, together with its French, German, Italian, and Japanese versions. The aim is to give precise definitions characterizing the dependability of computing systems. Dependability is first introduced as a global concept and a set of basic definitions is given. Those definitions are then commented, and supplemented, in the subsequent sections dealing respectively with the impairments to dependability (faults, errors, failures), the means for dependability (fault-prevention, fault-tolerance, fault-removal, fault-forecasting), and the attributes of dependability (reliability, availability, safety, security). The 116 definitions given throughout the text are recapitulated in a glossary, and a five-language cross-index is provided.

Vol. 4: *A. Avizienis, J.-C. Laprie (eds.)*
Dependable Computing for Critical Applications

1991. 88 figs. XIII, 431 pages.
Cloth DM 162,–, öS 1134,–
ISBN 3-211-82249-6

Vol. 3: *P.A. Lee, T. Anderson*
Fault Tolerance. Principles and Practice.

Third edition in preparation.

Vol. 2: U. Voges (ed.)
Software Diversity in Computerized Control Systems

1988. 41 figs. VII, 216 pages.
Cloth DM 75,–, öS 530,–
ISBN 3-211-82014-0

Vol. 1: *A. Avizienis, H. Kopetz, J.-C. Laprie (eds.)*
The Evolution of Fault-Tolerant Computing

In the Honor of William C.Carter
1987. 52 figs., 35 portraits and 1 frontispiece. X, 465 pages.
Cloth DM 118,–, öS 830,–
ISBN 3-211-81941-X

Prices are subject to change without notice.

Springer-Verlag Wien New York

Sachsenplatz 4-6, P.O.Box 89, A-1201 Wien · Heidelberger Platz 3, D-1000 Berlin 33
175 Fifth Avenue, New York, NY 10010, USA · 37-3, Hongo 3-chome, Bunkyo-ku, Tokyo 113